Mathematical Modelling Techniques

Mathematical Modelling Techniques

Rutherford Aris

University of Minnesota

DOVER PUBLICATIONS, INC.

New York

Acknowledgments

The material on pages 193–269 of this Dover edition originally appeared in the following sources:

—"Re, k and π: A Conversation on Some Aspects of Mathematical Modelling," *Appl. Math. Modelling*, Vol. 1 (1977), pp. 386–394. Reprinted with the permission of the publisher.

—"The Jail of Shape," *Chem. Eng. Commun.*, Vol. 24 (1983), pp. 167–181. Copyright © 1983 by Gordon and Breach, Science Publishers, Inc. Reprinted with the permission of the publisher.

—"The Mere Notion of a Model," *Mathematical Modelling*, Vol. 1 (1980), pp. 1–12. Copyright © 1980 Pergamon Press Ltd. Reprinted with kind permission from Pergamon Press Ltd, Headington Hill Hall, Oxford OX3 OBW, United Kingdom.

—"Ut Simulacrum, Poesis," *New Literary History*, Vol. 20 (1988–89), pp. 323–340. Reprinted with the permission of the publisher.

—"Manners Makyth Modellers," *Chemical Engineering Science*, Vol. 46, No. 7 (1991), pp. 1535–1544. Copyright © 1991 Pergamon Press plc. Reprinted with kind permission from Pergamon Press Ltd, Headington Hill Hall, Oxford OX3 OBW, United Kingdom.

—"How to Get the Most Out of an Equation without Really Trying," *Chemical Engineering Education*, Vol. 10 (1976), pp. 114–124. Reprinted with the permission of the publisher.

Copyright

Published in Canada by General Publishing Company, Ltd., 30 Lesmill Road, Don Mills, Toronto, Ontario.

Bibliographical Note

This Dover edition, first published in 1994, is an unabridged, slightly corrected republication of the work originally published by Pitman Publishing Limited, London, in 1978. The Dover edition has been expanded by the addition of six journal articles not included in the original edition.

Library of Congress Cataloging-in-Publication Data

Aris, Rutherford.
Mathematical modelling techniques / Rutherford Aris.
p. cm.
Originally published: London : Pitman, 1978.
"The Dover edition has been expanded by the addition of six journal articles not included in the original edition"—T.p. verso.
Includes bibliographical references and indexes.
ISBN 0-486-68131-9 (pbk.)
1. Mathematical models. I. Title.
QA401.A68 1994
511′.8—dc20 94-23196
CIP

Manufactured in the United States of America
Dover Publications, Inc., 31 East 2nd Street, Mineola, N.Y. 11501

To my friends at the California Institute of Technology, than whom there are no hosts more gracious nor colleagues more stimulating. Floreat Schola Artium Utiliorum Californiana.

Preface

"Par ma foi! il y a plus de quarante ans que
je dis de la prose sans que j'en susse rien,
et je vous suis le plus obligé du monde de
m'avoir appris cela."

M. Jourdain in Moliere's "Le Bourgeois
Gentilhomme" (Act. II Sc. IV)

The original title under which these notes were written--"Notes toward the definition of the craft of mathematical modelling"--was somewhat long-winded and perhaps, by reason of its allusion, a shade pretentious. It had however the merit of greater precision and conveyed the tentative spirit in which these notes are put forward for the criticism of a larger public. The whole activity of mathematical modelling has blossomed forth into such a multitude of areas in the last few years (witness a 1st International Conference with 2646 pages of Proceedings [199]) that there is indeed a need to define it in the sense of seeking out its boundary and exploring its interior as well as of discovering its structure and essential nature. The time is not yet ripe for a magisterial survey, which would in any case demand an abler pen than mine, but I believe it can be approached from the angle of craftsmanship. It is a commonplace in educational circles that it is comparatively easy to teach the method of solution of a standard mathematical equations, but much harder to communicate the ability to formulate the equations adequately and economically. With the notable exception of Lin and Segel's [223], and Haberman's book [213] and the papers of Hammersley [80,81,82] few publications pay much attention to the little things that the experienced mathematical modeller does, almost by instinct. It would therefore seem to be worthwhile to try and set down some of these notions in the interests of the craft and with the hope that it will stimulate further discussion and development. There is manifestly a danger here, for it may be only the MM. Jourdains who will be vastly excited to learn that they have been talking prose all their lives. Nevertheless I hope that some of my peers and betters will find the subject worthy of their attention.

Later iterations of this effort will demand a wealth of examples drawn from all branches of the physical and social sciences. In this first attempt I have chosen three physical examples to serve the illustration of many points. These examples--the packed bed, the chromatographic column and the stirred tank--are given in detail in the appendices. They are in some sense fold-out maps to the text though they cannot be presented as such. (Each has its own nomenclature which is listed at the end of its discussion; the nomenclature for other examples is introduced in situ.) These examples and those introduced at various points of the text are often connected with the mathematical theory of chemical reactors. I make no apology for this; the field is a rich one that has stimulated some of the work of the best applied mathematicians who have used a reactor like a stalking-horse under cover of which to shoot their wit. Its problems are challenging, yet from the modelling point of view they do not demand any great knowledge of chemistry or of engineering and so are accessible to all.

Many of the notions I have advanced here and the order I have tried to impose on the subject are quite tentative and I shall appreciate any comments and criticism. I have already benefitted from interaction with colleagues, both faculty and students, at Caltech and it is one of the virtues of

Pitman's Research Notes for Mathematics series that it quickly submits ideas to a wider public.

To the California Institute of Technology I am vastly indebted for a term as a Sherman Fairchild Distinguished Scholar in the fall of 1976, under conditions of such generous hospitality that the fruits of such a tenure can never be worthy of the opportunity. At the risk of overlooking someone, I would like to thank in particular (and in alphabetical order) Cohen, Gavalas, Keller, Pings, Seinfeld and Weinberg. Yolande Johnson did a splendid job of the first draft of these notes that was prepared at Caltech, being helped by Sharon ViGario in the last minute rush. The final version was typed by Shirley Tabis who met the exacting requirements of camera readiness with great skill and dispatch. I am most grateful to all of them.

Contents

1 What is a model?

ὦ Μένανδρε καὶ βίε
πότερος ἄρ' ὑμῶν πότερον ἀπεμιμήσατο

attr. to Aristophanes of Byzantium
Syrianus in Hermog. (Rabe ii 23)

(O Menander, o life, which of you has imitated the other?)

1.1 The idea of a mathematical model and its relationship to other uses of of the word.

In these notes the term 'mathematical model'--usually abbreviated to 'model'--will be used for any complete and consistent set of mathematical equations which is thought to correspond to some other entity, its prototype. The prototype may be a physical, biological, social, psychological or conceptual entity, perhaps even another mathematical model, though in detailed examples we shall be concerned with a few physico-chemical systems.

Being derived from 'modus' (a measure) the word 'model' implies a change of scale in its representation and only later in its history did it acquire the meaning of a type of design, as in Cromwell's New Model Army (1645). Still later (1788) came the complacent overtones of the exemplar that Gilbert was to use so effectively for his modern major general, while it is the first years of this century before fashion became so self-conscious as to claim its own models and make possible Kaplan's double entendre (see quotation at head of Ch. 5). In the sense that we are seeking a different scale of thought or mode of understanding we are using the word in its older meaning. However, the word model (without the adjective 'mathematical') has been and is used in a number of senses both by philosophers and scientists as merely glancing through the titles of the bibliography will suggest. Thus Apostel [7] distinguishes nine motivations underlying the use of models ranging from the replacement of a theory-less domain of facts by another for which a theory is known (e.g., network theory as a model for neurological phenomena) to the use of a model as a bridge between theory and observation. Suppes in the same volume [169] maintains that the logician's concept of a model is the same in the empirical sciences as in mathematics though the use to which they are put is different. The logician's definition he takes from Tarski [172] as: "a possible realization in which all valid sentences of a theory T are satisfied is called a model of T". This is a non-linguistic entity in which a theory is satisfied and Suppes draws attention to the confusion that can arise when model is used for the set of assumptions underlying a theory, i.e. the linguistic structure which is axiomatized. In our context this suggests that we might usefully distinguish between the prototype (i.e. the physical entity or system being modelled), the precursive assumptions or what the logicians call the theory of the model (i.e. the precise statement of the assumptions of axioms) and the model itself (i.e. the scheme of equations).

The idea of a change of scale which inheres in the notion of a model through its etymology can be variously interpreted. In so far as the prototype is a physical or natural object, the mathematical model represents a change on the scale of abstraction. Certain particularities will have been removed and simplifications made in obtaining the model. For this reason some hard-headed, practical-minded folk seem to regard the model as less "real" than the prototype. However from the logical point of view the prototype is in fact a realization in which the valid sentences of the mathematical model are to some degree satisfied. One could say that the prototype is a model of equations and the two enjoy the happy reciprocality of Menander and life.

The purpose for which a model is constructed should not be taken for granted but, at any rate initially, needs to be made explicit. Apostel (loc. cit.) recognizes this in his formalization of the modelling relationship R(S,P,M,T), which he describes as the subject S taking, in view of a purpose P, the entity M as a model for the prototype T. J. Maynard Smith [165] uses the notion of purpose to distinguish mathematical descriptions of ecological systems made for practical purposes from those whose purpose is theoretical. The former he calls 'simulations' and points out that their value increases with the amount of particular detail that they incorporate. Thus in trying to predict the population of a pest the peculiarities of its propagation and predilections of its predators would be incorporated in the model with all the specific detail that could be mustered. But ecological theory also seeks to make general statements about the population growth that will discern the broad influence of the several factors that come into play. The mathematical descriptions that serve such theoretical purposes should include as little detail as possible but preserve the broad outline of the problem. These descriptions are called 'models' by Smith, who also comments on a remark of Levins [114] that the valuable results from such models are the indications, not of what is common to all species or systems, but of the differences between species of systems.

Hesse [92] in her excellent little monograph "Models and Analogies in Science" distinguished two basic meanings of the word 'model' as it is used in physics and Leatherdale in a very comprehensive discussion of "The Role of Analogy, Model and Metaphor in Science" has at least four. They stem from the methods of "physical analogy" introduced by Kelvin and Maxwell who used the partial resemblance between the laws of two sciences to make one serve as illustrator of the other. In the hands of 19th century English physicists these often took the form of the mechanical analogues that evoked Duhem's famous passage of Gallic ire and irony. Duhem [56] had in mind that a physical theory should be a purely deductive structure from a small number of rather general hypotheses, but Campbell [41] claimed that this logical consistency was not enough and that links to or analogies with already established laws must be maintained. Leatherdale's four types are the formal and informal variants of Hesse's two. Her "model_1" is a copy, albeit imperfect, with certain features that are positively analogous and certain which are neutral but shorn of all features which are known to be negatively analogous, i.e. definitely dissimilar to the prototype. Her "model_2" is the copy with all its features, good, bad and indifferent. Thus billiard balls in motion, colored and shiny, are a model_2 for kinetic theory, whilst billiard balls in motion obeying perfectly the laws of mechanics but bereft of their colour, shine and all other non-molecular properties constitute a model_1. It is the natural analogies (i.e. the features as yet of unknown relevance) that are regarded by Campbell as the growing points of a theory. In these terms a mathematical model would presumably be a formal model_1.

Brodbeck [35], in the context of the social sciences, stresses the aspect of isomorphism and reciprocality when she defines a model by saying that if the laws of one theory have the same form as the laws of another theory, then one may be said to be a model for the other. There remains, of course, the problem of determining whether the two sets of laws are isomorphic.

Brodbeck further distinguishes between two empirical theories as models one of the other and the situation when one theory is an "arithmetical structure". She then goes on to describe three meanings of the term mathematical model according as the modelling theory is (a) any quantified empirical theory, (b) an arithmetic structure or (c) a mere formalization in which descriptive terms are given symbols in the attempt to lay bare the axioms or otherwise to examine the structure of the theory. If arithmetical is interpreted with suitable breadth we are clearly concerned in these notes with sense (b).

It is obviously inappropriate in the present context to try to survey all the senses in which the word has been used, among which there is no lack of confusion. A more formal version of the definition of a (mathematical) model that we started with might be as follows: a system of equations, Σ, is said to be a model of the prototypical system, S, if it is formulated to express the laws of S and its solution is intended to represent some aspect of the behavior of S. This is vague enough in all conscience, but the isomorphism is never exact and we deny the name of modelling to the less successful efforts of the game. Rather, we should try and find out what constitutes a good or bad model.

It scarcely needs to be added that we shall not raise the old red herring about the model being less "real" than the prototype. Tolkien [178] has reminded us of the failure of the expression "real life" to live up to academic standards. "The notion", he remarks, "that motor cars are more 'alive' than, say, centaurs or dragons is curious; that they are more 'real' than, say, horses is pathetically absurd".

The mention of reality leads me to add that by far the most enlightening discussion of models I have found is in Harré's excellent introduction to the philosophy of science, "The principles of scientific thinking" [37]. He writes from a realist point of view which eschews simplifications and attempts to present a theory of science based on the actual complexity of scientific theory and practice; he regards the alternative traditions of conventialism and positivism as vitiated by the attempt to force the description of scientific intuition and rationality into the deductivist mould. Model building becomes an essential step in the construction of a theory. I shall not attempt to summarize the argument of his second chapter, which demands careful and considered reading, but it may be useful to mention one or two of the distinctions he makes. He starts with the notion of a sentential model in which one set of sentences T is a model of (or with respect to) another set of sentences S if for each statement t of T there is a corresponding statement s of S such that s is true whenever t is acceptable and t is unacceptable whenever s is false. If T and S are descriptions of two systems M and N and T is a sentential model of S, then M is an iconic model of N. He recognizes that in mathematics the word is used in both ways: model theory is clearly a sentential model within mathematical logic, but we often conceive sets of objects, real or imaginary, which are described mathematically. The latter is an iconic model and the equations a sentential model of the sentences describing the set of objects. Harré goes on to distinguish between the subject and source of a model. The former is whatever the model is a model of, the latter what it is based on; for example elementary kinetic theory gives models of a gas (subject) based on the mechanics of particles (source). Homeomorphs are models in which the source and subject are the same as in a mechanical scale model. When source and subject are not the same, as with the English tubes and beads that amazed Duhem so much (cf. Sec. 2.1), Harré speaks of paramorphs. He goes on to discuss the taxonomy of models and to show how they are incorporated into the construction of theories first by the creation of a paramorph and then by supposing that it provides a hypothetical mechanism. This process evokes existential hypotheses and raises such questions as the degree of abstraction that can be tolerated leading into a full-scale discussion of the formation of scientific theories. Clearly mathematical modelling in the sense in which we are here discussing it is a small part of this much larger design.

1.2 Relations between models with respect to origins.

It seems well to use the term model for any set of equations that under certain conditions and for a certain purpose provide an adequate description of a physical system. But, if we do this, we must distinguish the kinds of relationships that can obtain between different models of the same process. (This approach seems more useful than to talk of models and sub-models, since the relations are more varied and mixed than can be compasses by this nomenclature). It is of the first moment to recognize that models do not exist in isolation and that, though they may at times be considered in their own terms, models are never fully understood except in relation to other members of the family to which they belong.

One type of relationship can be seen in the packed bed example, the full details of which are given in Appendix A. The physical system is that of a cylindrical tube packed with spherical particles and our purpose is to model the longitudinal dispersion phenomenon. By this we mean that if a sharp pulse of some tracer is put into the stream flowing through a packed bed it emerges as a broad peak at the far end of the bed, showing that some molecules of the tracer move faster through the bed than others and that the sharp peak of tracer is dispersed.

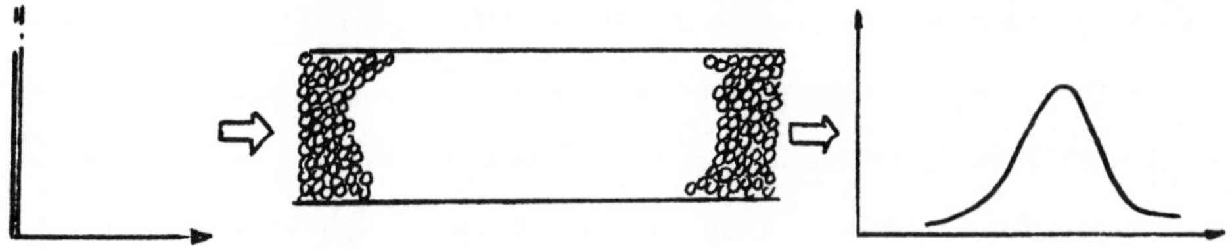

This is the physical system P and it is amenable to modelling in various ways. The most obvious one is to write down the equation of continuity, the Navier-Stokes equation and the diffusion equation with their several boundary conditions (eqns. A1-7). This model, which we will call Π_1, is admirably complete and founded on the fewest and most impeccable assumptions, but, for two reasons, it is not a very useful model. In the first place, if the actual geometry of a given packed bed could be used the results would be peculiar to that bed, making it a good simulation but a bad model in the senses of Smith. Secondly, even if the geometry were standardized (and this presents its own difficulties) to, say, a cubic array of spheres the resulting equations would present ferocious difficulties to computation and the model would probably remain barren of results. If the models with standardized and peculiar geometries are denoted by Π_1' and Π_1'' respectively they are clearly distinguishable but very closely related--in fact almost "non-identical twins".

The second way in which we might try to model P is to say that the same sort of dispersion is experienced in a much simpler system, namely that of plug or uniform flow through a tube with a longitudinal diffusion coefficient. If we call this modified prototype P_2 we can easily derive a partial differential equation Π_2, which is much simpler than those of Π_1 (see eqns. A8-12). There is no immediate connection between Π_1 and Π_2 though we can imagine that some sort of averaging of the Navier-Stokes equations over the cross-section of the bed would lead to the plug flow approximation.

On the other hand, we might make something of the fact that in a packed bed the space between particles makes a natural cavity whilst the interstices narrow where the particles touch and the fluid can be thought of as jetting through into the cavity space. In this rather crude sense the packed bed as a sequence of little stirred tanks gives us a modified prototype, say P_3, which can be modelled. To avoid Suppes' criticism, we do not say that P_3 is a model of P, though recognizing that it is popularly called the "cell model" of the packed bed. The model of P_3 (and therefore of P) consists N ordinary differential equations for the time-varying concentrations in the N stirred tanks of P_3. This will be denoted by Π_3 and the equations are numbered A13-16. There is no obvious connection between Π_1 and Π_3 or between Π_2 and Π_3.

A fourth way of modelling the system P would be to regard the system as a stochastic one in which a tracer molecule had at each step in time the options of either moving forward with the stream or of being caught in an eddy and remaining essentially in the same place. This modification of the prototype, say P_4, leads to Π_4 and the equations A25-26. Again there is no immediate or obvious connection between Π_4 and the preceding models. The relationship of these models is expressed in the diagram.

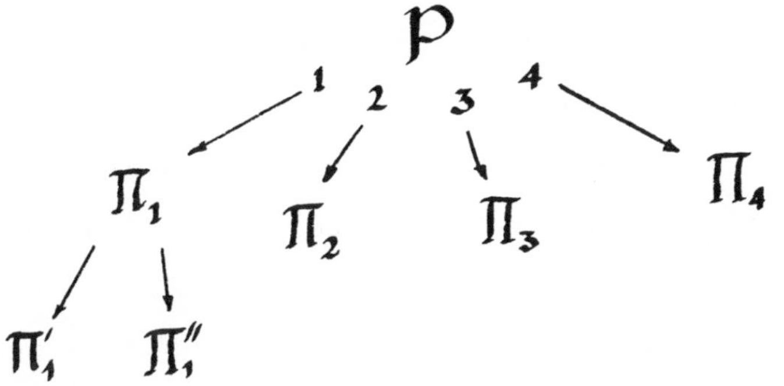

The models $\Pi_1,\ldots\Pi_4$ are best described as cognate models since they appear to be siblings of the same parent system.

A rather different relationship obtains between the models $\Sigma_1,\ldots,\Sigma_6$ of the stirred tank S described in Appendix C. In these Σ_1 is the full set of ordinary and partial differential equations obtained by making mass balances for each of the S species and energy balances on the contents, the wall, and the cooling system of the reactor (see pp. 152-164 for details and particularly pp. 153 and 154 for the hypotheses). This gives S+2 ordinary differential equations and a parabolic partial differential equation. This system, Σ_1, (eqns. C1-6) is again of considerable complexity, but less difficult of calculation than P_1. Σ_2 is the steady state version of these equations obtained simply by deleting all time derivatives and with them the initial conditions. It thus consists of non-differential equations coupled to an elliptic differential equation. If the assumption is made that the wall of the reactor is thin, (hypothesis H_6) the elliptic equation can be solved quite easily and Σ_3 then consists entirely of algebraic equations, C9 and 10. (We will call equations that are not differential equations 'algebraic' even though they may contain transcendental functions).

To reach the model Σ_4 we return to the full transient model Σ_1 and assume that the conductivity of the wall is very high (hypothesis H_7). Then the parabolic partial differential equation can be replaced by an ordinary differential equation for the mean wall temperature and Σ_4 consists wholly of ordinary differential equations one for each reacting species and one for each of the reactor, wall and coolant temperatures. (eqns. C1, 2, 6, 11) If H_6 and H_8, the hypotheses that assert that the wall is thin and of negligible heat capacity, are imposed instead we have one fewer equation and the model Σ_5 (eqns. C1, 13, 14).

The final model, Σ_6, is written for a special case of some importance, both historically and logically. It is a case that has dominated the development of stirred tank reactor analysis, as seen in the papers of van Heerden [253], Amundson and Bilous [258], Amundson and Aris [5] and Uppal, Ray and Poore [183,184]. Logically it can be defended as the simplest case in which the essentially nonlinearity of the nonisothermal behavior comes to light. Thus the added hypotheses of instantaneous cooling action (H_9) and restriction to a single irreversible, first-order reaction (H_{10}) allow the system to be immediately reduced to a pair of ordinary differential equations (C15 and 16).

The relationship of those models is represented by the diagram:

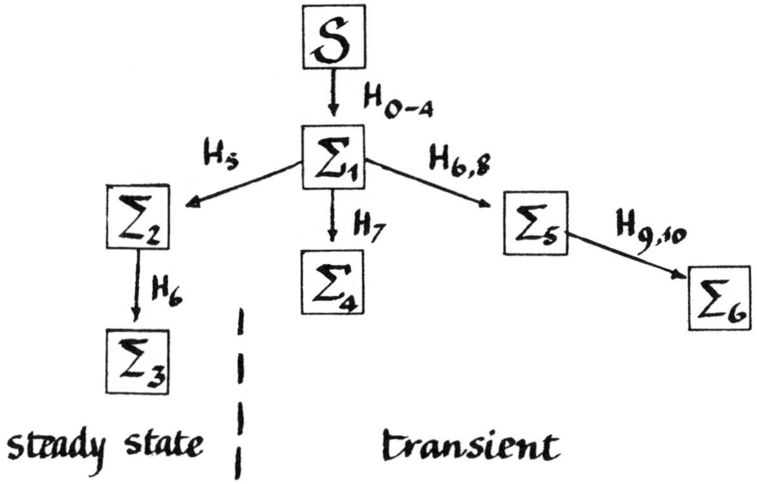

It makes more sense in this case to speak of $\Sigma_2, \ldots \Sigma_6$ as derived from (rather than cognate with) Σ_1 since no modification of S is involved and the later models can be obtained from Σ_1 by letting certain parameters take on limiting values. Steady-state models are worth singling out as of particular importance. They can be obtained formally as a limiting case introducing an artificial parameter multiplying each time derivative and letting this go to zero. But this should be distinguished from a pseudo-steady state hypothesis such as H_9 in which a physical parameter is very small and is allowed to take its limiting value of zero. Thus we have a clone of models in which Σ_2, Σ_4 and Σ_5 are immediately derived from Σ_1 while Σ_3 and Σ_6 are derived from Σ_2 and Σ_5.

The reader is left to justify the diagram given on p. 149 for the relationship of the models of the chromatograph given in Appendix B. They are not predominantly cognate, as are the Π_i, or derived, like the Σ_j, but appear to be a mixture of both. No new relationships seem to be introduced however.

The model of a model represents a relationship which is a little different from that of cognate models or from the idea of a derived model. This kind of modelling arises when the first model is so complicated, either in the form of the equations or the number of parameters, that it seems that better insight can be gained by quite drastic simplification. Such is the case when Burger's equation $u_t+uu_x = \nu u_{xx}$ is used to get preliminary insight into the nature of turbulence. It is not claimed that the physical system corresponds exactly to this equation, though it may be that an artificial one could be constructed. But the model of the model has its validity in so far as it extracts some important feature of the first model with a form in which it can be analyzed more easily. The relationship is more like:

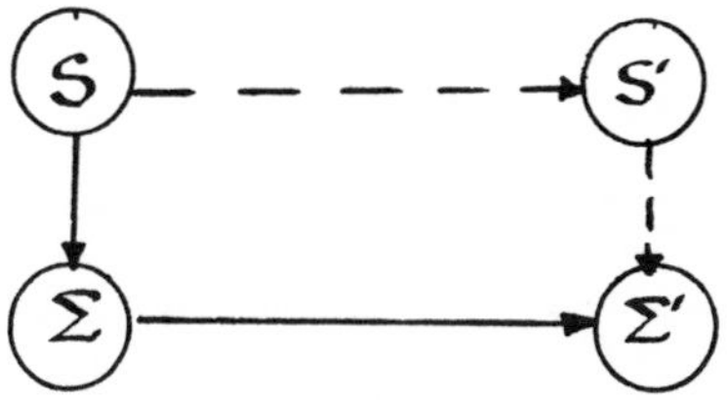

where Σ is the first model and Σ' the model of it. The relationships indicated by the dotted line may or may not exist and even when they exist they may or may not be worth attending to.

Another example is the Lorentz equations

$$\dot{x} = -\sigma(x-y)$$

$$\dot{y} = -xz + rx - y$$

$$\dot{z} = xy - bz$$

which have a tenuous connection with meteorology, whose equations are well developed but vastly complicated; they are perhaps best regarded as a model of the meteorological equations. How profitable, and indeed fascinating, the study of them may be is to be seen from Lorentz' [119] and Marsden and McCracken's [128] treatments of them.

In some cases the difficulties with Σ, the model of the physical system S, lead the investigator to consider a simplified system, S', and construct a model, Σ', of it. In this case the connection between Σ and Σ' may not be of interest and the situation is

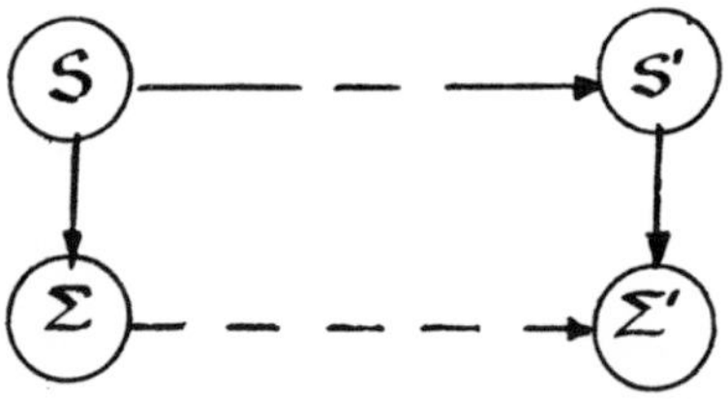

Such a case is the famous "Brusselator" where the unlikely reaction mechanism $A \to X$, $B+X \to Y+D$, $2X+Y \to 3X$, $X \to C$ was inspired by the Belousov-Zhabotinsky reaction. It has much intrinsic interest and served to bring to light some important phenomena (see e.g. Nicolis and Portnow [135] Lavenda, Nicolis and Herschkowitz-Kaufmann [109] or the introductory summary in [14]. But it has little to do with the Belousov-Zhabotinsky reaction itself, the "Oregonator" and other mechanisms being much more direct attempts to give precursors of a model for this. The reader should refer to the excellent little monograph of Tyson [182] and the references given there, particularly the papers of Kopell and Howard, Murray, Othmer and Winfree [see cf. 219,254].

Another example stimulated by the need to reduce both the number of parameters and the number of equations in Aris and Schruben's [17] simplification of Amundson and Liu's [6] packed bed equations. Defining the heights of transfer units for mass and heat between the particles and intersticial fluid as H_g and H_h, respectively, Amundson and Liu had written

$$\frac{\partial p}{\partial x} + \frac{1}{u}\frac{\partial p}{\partial \Theta} = \frac{p_p - p}{H_g} \tag{1.1}$$

$$\frac{\partial t}{\partial x} + \frac{1}{u}\frac{\partial t}{\partial \Theta} = \frac{t_p - t}{H_h} \tag{1.2}$$

where p = partial pressure of the reactant in the fluid,

p_p = partial pressure in the particle where reaction takes place.

t = temperature of fluid,

t_p = temperature of particle,

x = distance from inlet

Θ = time

These equations were derived from mass and heat balances in the fluid and similar handling of the particles, assuming a first order reaction rate $k(t_p)p_p$, gave

$$L_g \frac{\partial p_p}{\partial \Theta} = p - p_p - M_g k(t_p)p_p \tag{1.3}$$

$$L_h \frac{\partial t_p}{\partial \Theta} = t - t_p - M_h k(t_p)p_p \tag{1.4}$$

where $L_g, \ldots M_h$ are further combinations of the flow rate, particle size, density of fluid, etc., the details of which need not concern us here. If we make p, p_p dimensionless by dividing by p^*, t and t_p by t^*, x with x^* and θ with Θ^* we are left with five parameters $u\theta^*/x^*, H_g/x^*, H_h/x^*, L_g/\Theta^*, L_h/\Theta^*$ and the additional parameters in $M_g k(t_p)$

and $M_h k(t_p)p^*/t^*$ of which there are three. Of these eight only four can be eliminated by the choice of x^*, Θ^*, p^* and t^*; for example, if $x^* = H_g$, $\Theta^* = H_g/u$, $p^* = t^*/M_hA$, $t^* = E/R$ where $k(t_p) = A\ \exp-(E/RT_p)$, we are still left with L_gu/H_g, L_hu/H_g, H_h/H_g and M_gA. In the steady state the first two disappear, being multipliers of the time derivatives.

If we look at the steady state equations we have

$$H_g \frac{dp}{dx} = p_p - p = -M_g k(t_p) p_p \tag{1.5}$$

$$H_g \frac{dt}{dx} = (H_g/H_h)(t_p - t) = mM_g k(t_p) p_p \tag{1.6}$$

where $m = M_h H_g / M_g H_h$. Thus from the two end terms of the equations

$$H_g \frac{d}{dx} [mp + t] = 0 \tag{1.7}$$

or

$$p = p_e - (t - t_e)/m \tag{1.8}$$

where p_e and t_e are the entrance values. This last equation merely expresses the fact that the bed is adiabatic. Also from the last two terms of eqn. (1.6).

$$t_p - t = \frac{M_h p}{M_g} \frac{M_g k(t_p)}{1 + M_g k(t_p)} \tag{1.9}$$

Using eqn. (1.8)

$$Q_1 \equiv \frac{M_g}{M_h} \frac{t_p - t}{p_e + (t_e/m) - (t/m)} = \frac{M_g k(t_p)}{1 + M_g k(t_p)} \equiv Q_2 \tag{1.10}$$

The right hand side of this equation is an S-shaped curve depending only on the two parameters in $M_g k(t_p)$. The left hand side varies from point to point in the bed since it depends on t. However it represents a family of straight lines all passing through the point $(t_e + mp_e, H_g/H_h)$. For any t at position x, t_p can

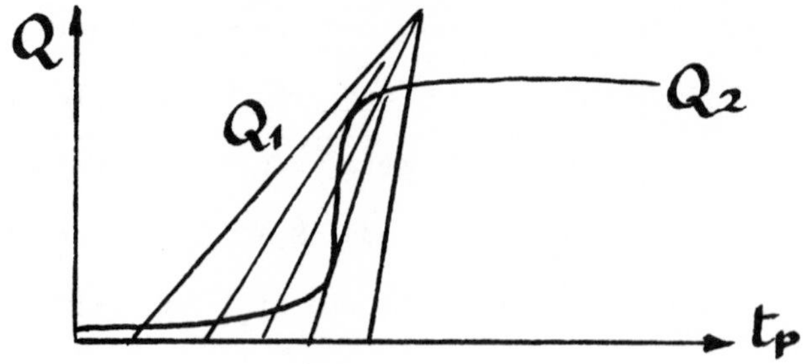

be found by solving eqn. (1.10) for t_p and eqn. (1.5) for p_p. Then with p given by eqn. (1.8), there is a single equation

$$\frac{dt}{dx} = f(t) \tag{1.11}$$

to solve for $t(x)$ and this can even be done by quadrature. However the diagram shows that the solution of eqn. (1.11) may be multivalued and this is the origin of the multiple profiles that are possible in a packed bed. Eigenburger [59,60] showed that this continuum of steady states is reduced to a single state if heat conduction between particles is allowed, but the same phenomenon can be found in isothermal beds with more complicated kinetics.

Because of the number of parameters there seems no possibility of getting a comprehensive view of the system, though it is to be noted that the multiplicity of the solutions of eqn. (1.10) is determined by the study of the stirred tank in Sec. 5.2, if

$$\alpha = M_g A, \quad \zeta = RM_h p/M_g E, \quad v = Rt/E.$$

This suggests that a simpler model should be constructed that would incorporate this feature of multiplicity with the fewest parameters possible. This was the motivation of Aris and Schruben [17].

Note first that the intermediate intersection can be disregarded since it is known always to be unstable. This suggests that the sigmoid curve should be replaced by a step function i.e.

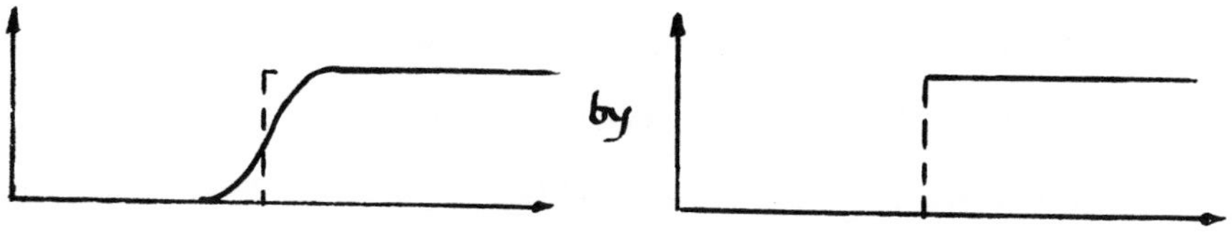

Also it seems desirable to reduce the number of equations to two. This can be done if we suppose that the wall of a tube generates heat at a rate Q by means of an exothermic reaction of order zero which is triggered at a critical temperature T_c. This heat from the wall, whose temperature at position x and time t is denoted by $W(x,t)$ is transferred to the fluid flowing in the tube, whose temperature is $T(x,t)$, and we have the equations

$$c_w a_w \frac{\partial W}{\partial t} = h_p(T-W) + QH(W-T_c)$$

$$c_f a_f \left[\frac{\partial T}{\partial t} + V\frac{\partial T}{\partial x}\right] = h_p(W-T)$$

where c_w, c_f are the heat capacities of wall and fluid,
a_w, a_f are their areas,
p is the inner perimeter of the tube,
h is the heat transfer coefficient,
H is Heaviside's step function,
and V is the velocity of the fluid.

The substitutions

$$v = h_p(T-T_c)/Q, \quad w = h_p(W-T_c)/Q$$

$$\xi = h_p x/Vc_f a_f, \quad \tau = h_p t/c_f a_f$$

$$\omega = c_w a_w/c_f a_f$$

give

$$\frac{\partial v}{\partial \tau} + \frac{\partial v}{\partial \xi} = w-v$$

$$\omega\frac{\partial w}{\partial \tau} = v-w + H(w)$$

These equations have only one parameter and the solution of the nonlinear equation $v-w + H(w) = 0$ is immediate. This allowed the full spectrum of possible solutions to be surveyed and some observations on the transients to be made that had a bearing on the more complicated system. They also have an important bearing on the behavior of the monolithic reactor.

1.3 Relations between models with respect to purpose and conditions.

The relationship between models is not only an intrinsic matter of mathematical genealogy but must be viewed also in the perspective of the purpose of the model and the conditions under which it is to be used. For example, the steady state models of the stirred tank, Σ_2 and Σ_3, are quite unfitted to the purposes of control whatever the conditions may be. They are adapted to the purpose of steady state design however. Σ_2 may be demanded by the conditions of thick walls or poor heat conduction but Σ_3 suffices otherwise. This is indicated in part (a) of the figure below, where the boundaries of

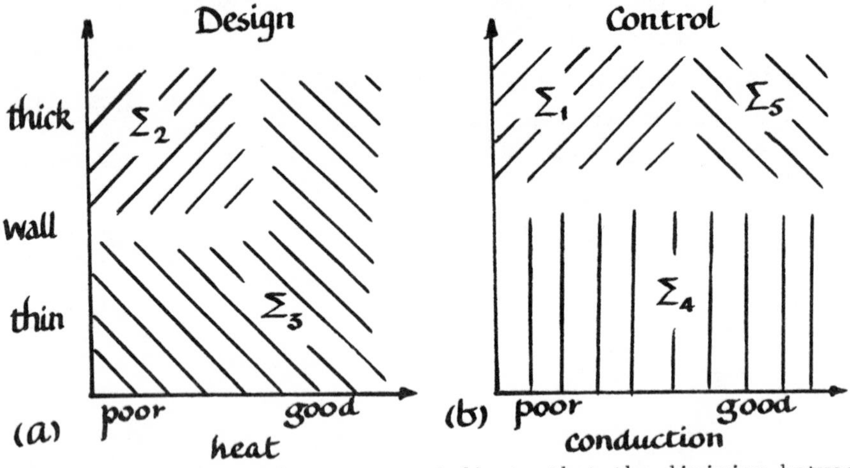

the regions are deliberately vague to indicate that the division between the regions of applicability is not a sharp one. If the purpose is control then Σ_1, the most complicated model, may be demanded by the same conditions that demand Σ_2 for the steady state. Σ_4 is valid if the wall is thin whatever its conductivity and Σ_5 is a simplification made possible by the good conductivity of the wall even when it is thick. Another way of representing the interaction of conditions and purposes with the type of model is shown in the next figure. Here the degree of sophistication of the model ranges from mere algebraic equations at the lowest level to the coupled partial and ordinary differential equations of Σ_1.

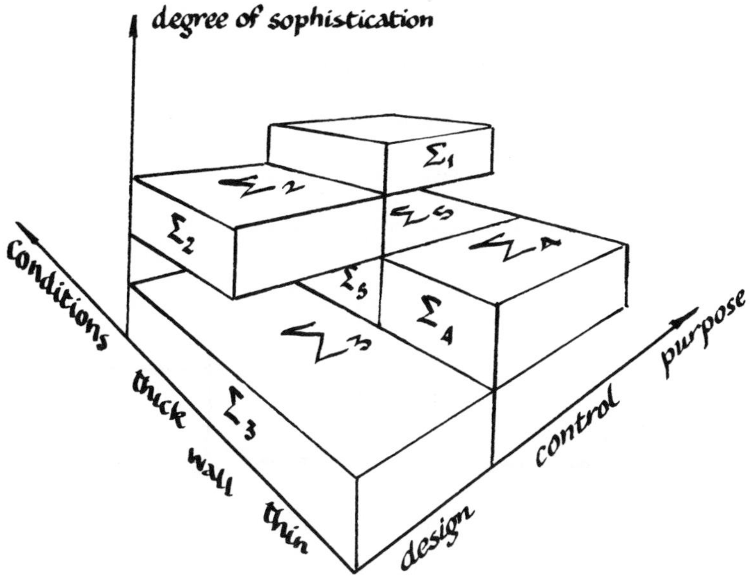

Sometimes the regions of applicability of models can be delineated more exactly by setting up a certain standard of accuracy. Though this is admittedly arbitrary it sets the stage for the entrances and exits of the models. An example is contained in the early work of Gill and his colleagues on Taylor diffusion [74,75] which may be illustrated by the Δ-models of Appendix B. The problem is to describe the movement of solute as the solvent passes in laminar flow through a long tube. The combined influence of molecular diffusion and convection softens an initially sharp front, for, thanks to the parabolic flow profile, tracer molecules at the center of the tube are taken ahead of the pack by the fast central streams but, then finding themselves in a region of low tracer concentration, can diffuse outwards to the slower streams and so slow down. The mean concentration of tracer is thus diminished in an error-function fashion about a point moving with the mean speed of the stream. In fact this mechanism of dispersion, first elucidated by Taylor in 1953 [174], shows that the higher the diffusion coefficient the smaller the longitudinal dispersion, since lateral diffusion will immediately negate the effects of the flow profile. But when the diffusion is isotropic a high lateral molecular diffusion implies a high longitudinal dispersion and there is a point at which total effect is minimum. The purpose of the model is to account for the advancing wave of solute as represented by its mean value. This is zero initially and in a general way is as shown here:

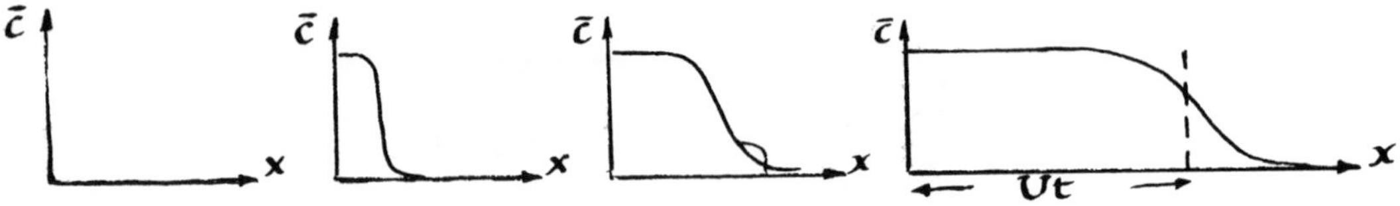

The models, described in detail in Appendix B, are:

Δ_2 the full parabolic partial differential equations to be solved for the concentration c as a function of x, distance from inlet, r, distance from tube axis, and time, t. This solution is averaged over the cross section to give $\bar{c}(x,t)$.

Δ_4 the equation for plug flow at the mean speed of the stream with an equivalent longitudinal diffusion coefficient D_e are solved for the mean concentration directly. D_e is related to D, the molecular diffusion coefficient, a, the tube radius and U, the mean velocity.

Δ_4' $D_e = a^2U^2/48D$

Δ_4'' $D_e = D + A^2U^2/48D$

Δ_4''' $D_e = D$

Δ_5 A pure convection model with no consideration of diffusion

Δ_6 An empirical fit of D_e between the solutions of Δ_2 and Δ_4.

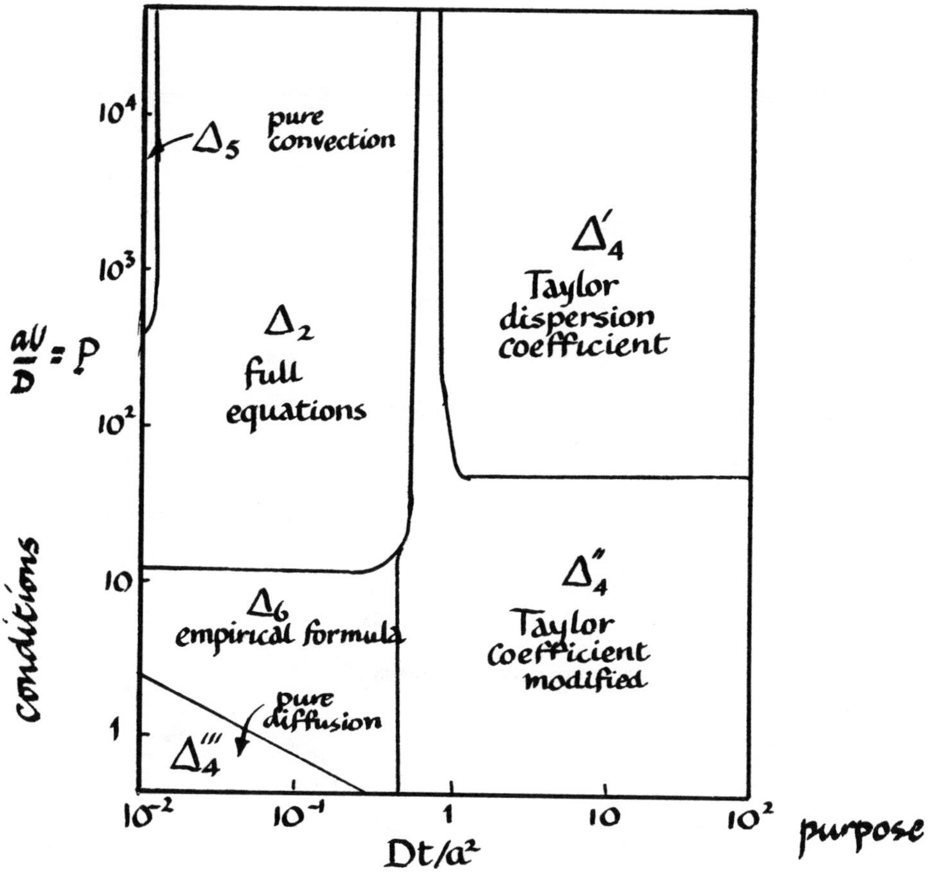

If the purpose is to provide a sufficiently accurate value of $\bar{c}$ for a period of time t we may take one axis to be the dimensionless time $\tau = Dt/a^2$. The conditions are represented by the Peclet number $P = aU/D$. Gill and his colleagues [74] mapped the τ-P space as shown above. Thus, for example, the full equations have to be solved in the region Δ_2, but the Taylor diffusion coefficient $a^2U^2/48D$ in the plug flow model suffices in the region Δ_4'. The arrangement of the Δ_4-regions is reasonable, for the modification Δ_4'' (cf. [10]) can be written $D[1 + P^2/48]$. Thus when P is large we can neglect the 1 in the bracket giving Δ_4', whereas, when P is small, $P^2/48$ is negligible. Δ_6 is an empirical bridge between Δ_4'' and Δ_4'''.

1.4 How should a model be judged?

We shall have more to say of the detailed evaluation of models in the last chapter, but it will be useful to make a few points in a preliminary way. Clearly, once a model has been set up it has a life of its own and its equations have their own intrinsic interest. However most models are indissolubly bound to their origins and cannot be viewed in isolation from their background any more than they can ignore their relatives. The relation is a dynamic one calling for continual interaction if either conceptual progress or actual understanding is to be gained. It may be envisioned in the following way:

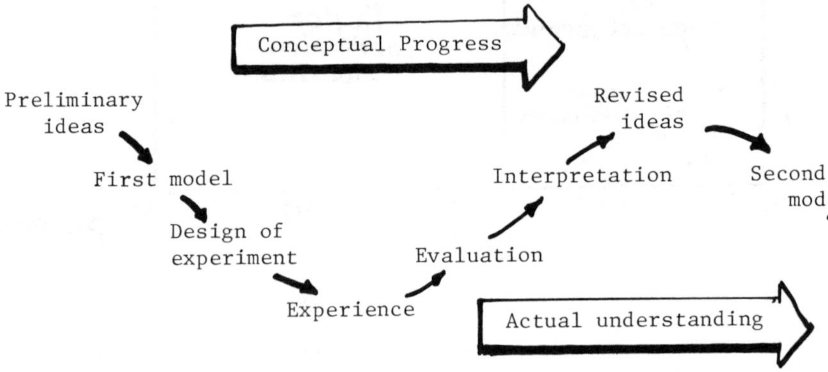

A certain primacy belongs to theory since it is impossible to design a good experiment without some theoretical vision and the more precise the theory the more decisive the experiment can be. If conceptual progress is to go hand in hand with the understanding of an actual situation there must be this intercourse between the system, S, and its family of models, Σ, though the distinction that Smith has made between model and simulation is a valid one. It might be represented by a serpentine progress that tends to emphasize one level of the other, i.e.:

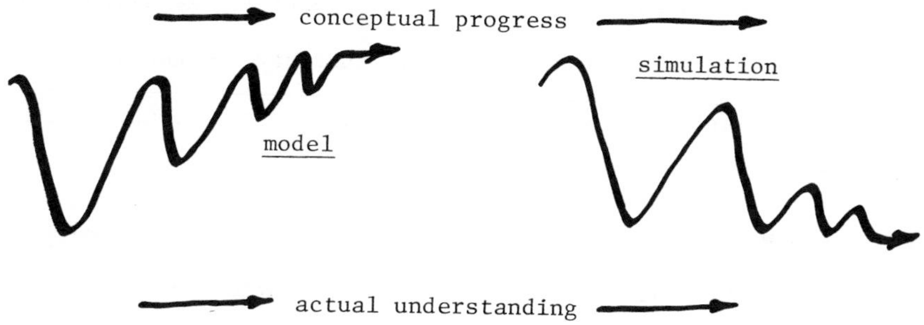

The revision of ideas and development of models is not necessarily in the direction of greater complexity or an increasing number of parameters. Progress may be toward simplification and the reduction of the number of adjustable constants since it is often said that you can fit an elephant with five constants, though Wei [189] has shown that the fit may not be spectacular. While it is certainly convincing if a complicated situation can be represented by the adjustment of very few constants, care must be taken to see that it is not purchased by a less obvious accommodation elsewhere, as when it was said of a certain theorist that his work required no adjustable constants but completely pliable hypotheses. Nevertheless part of the judgment of a model will lie in whether its constants can be found from independent sources and combined to give a convincing picture in the interactive situation. Thus we have a high degree of confidence in a model of a reactor embodying such complicating features as diffusion within catalyst pellets if the kinetics of the reaction, diffusivities and catalyst properties can all be determined independently and give, in the model, a recognizable behavior. We have less confidence if we adjust the parameters in the model as a whole, for example by a least square fit of outputs, without any interior understanding. Similarly at the interior level in determining, say, the kinetics, we have much more confidence in a kinetic expression that is based on an independently confirmed mechanism than in an overall kinetic expression that lacks this insight. Of course, the exigencies of a particular situation may force us to be content with less than the best.

There is however a phenomenon of current interest that raises deeper questions about our ability to compare the implications of a model with experience. This is the class of solutions of quite simple equations which are said to be chaotic in their behavior. The locus classicus in which this behavior may be observed is the difference equation

$$x_{n+1} = \lambda x_n(1-x_n)$$

--an innocent enough starting point for anyone. If $0 \le \lambda \le 4$ the transformation maps the interval [0,1] into itself. The origin $x = 0$ is always a fixed point and indeed is the only fixed point until λ exceeds 1 when $1-\lambda^{-1}$ is also a fixed point. At $\lambda = 1$ the origin becomes unstable but the new fixed point is stable and remains so until $\lambda = \lambda_2 \equiv 3$. At this point two stable solutions of period 2 (i.e. for which $x_{n+2} = x_n$) appear and are stable. But not for long, since when λ reaches $\lambda_4 = 1+\sqrt{6} = 3.45$ these become unstable and spawn four solutions of period 4 which are, at first, stable. This process of binary fission becomes increasingly frequent as λ increases and in fact the sequence of points $\lambda_2, \lambda_4, \lambda_8, \ldots$ at which the 2^n stable cycles of period 2^n appear has a limit point $\lambda_c = 3.57$. Beyond this point there are a countable infinite number of unstable periodic orbits and also an infinite number of solutions which are in no sense periodic, that is they are neither periodic nor asymptotic to a periodic solution. At $\lambda = 3.68$ a long periodic solution of odd length appears and by the time $\lambda = \lambda_3 \equiv 3.83$ a cycle of period 3 is seen. At this point cycles of all periods are present some of which are stable. In fact cycles of period 3 bifurcate into those of period 6, which in turn go to those of period 12 and again the values of λ at which these bifurcations take place have a limit point ($\lambda = 3.85$) short of $\lambda = 4$. Indeed Li and Yorke [115,222] have shown that once there is a cycle of period 3 there must be cycles of all periods as well as strictly non-periodic solutions--a situation aptly described as chaotic.

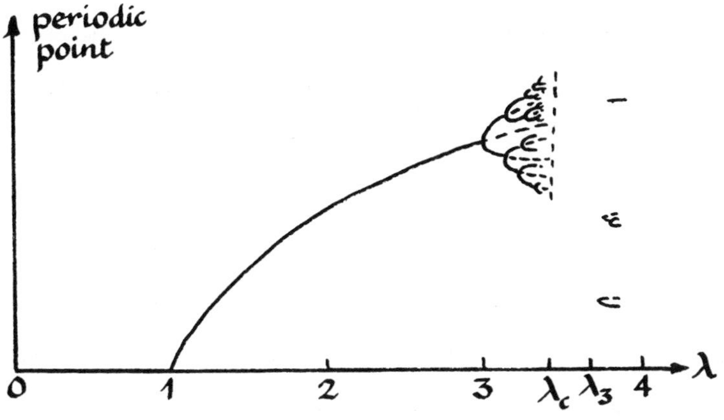

This remarkable situation is not peculiar to the expression $\lambda x(1-x)$ but is generic for all functions with a hump and a parameter such as λ by which it can be 'tuned'. It is described with admirable clarity by May, Oster and Guckenheimer in various places [128,129,130,211]. It is a feature shared by many allegedly simple systems as the Lorentz equations [119] and the examples of Rössler [157] show. In itself it points up the value of even the simplest models for here a simple generic situation has opened our eyes to a new type of behavior of the solution which may well reflect the irregularities experienced in nature. It differs from a random process in the following sense. In the random process the attempt to predict future states is limited by the range of the correlation of the random process, whereas in a chaotic process it is limited by the accuracy with which the initial conditions can be determined. This is the case because arbitrarily near to the initial point of any solution there are infinitely many initial points that will give solutions that ultimately diverge completely from the first solution. But this raises the question of whether the matching of the results of a computation with experience can ever be trusted. Such questions as: does a mismatch constitute an adverse reflection on the model or is it only the result of a failure to find the initial conditions with sufficient accuracy? could a chaotic solution be distinguished from one of a very long period? even if such a distinction could be made would it matter? how can two models ever be compared if their solutions are both chaotic? These and other questions are as yet unanswered but are relevant to the question of evaluating a model.

Different types of model call for different modes of evaluation. Bush and Mosteller [38] present an interesting comparison of eight statistical models of a learning process and in doing so examine various criteria. They reject the likelihood ratio, for example, as an adequate tool for discrimination. Though it is a convenient summary of the fit of the model to the data it is often difficult to compute and obscures the peculiar strengths and weaknesses of a particular model, failing to suggest why the model is inadequate. Moreover, it may be sensitive to uninteresting differences between the model and the experimental set-up. Instead they use a number of different statistics (e.g. trials before first avoidance, trials before last shock, etc.) to judge the match between the data and the learning sequences calculated from each model.

2 The different types of model

"Let us take them in order. The first is the taste,
which is meagre and hollow, but crisp: ...."

C. L. Dodgson. The Hunting of the Snark.
Fit 2. St. 16.

This chapter is little more than an annotated list of various types of model that are in common use. It would be pleasant, but unrealistic, to think that the author could produce a sequence of sparkling little essays on the historical origins, the current status, the manifold areas of application and the virtues and vices of each. No doubt this would be the very model of such a chapter but, for the moment, this rather superficial survey must suffice as a reminder of the variety of tools that the modeller should have in his bag.

2.1 Verbal models and mechanical analogies.

This type of model, referring to one couched in the language of everyday discourse rather than in the language of mathematics, is in the strict sense outwith our present policies, but it should nevertheless be mentioned since it may well have some of the characteristics of a mathematical model. Such models might be called "soft" models without being pejorative for they are useful without being "hard" in the sense of having quantitative predictions or sharply defined concepts. Toffler [88, p. 274] points out that in the social context it is virtually impossible to make absolutely explicit all the assumptions of a hard model and that models can thus be "implicitly soft." On the other hand there are "explicitly soft" models and Toffler points to the novelists who present verbal models of society often with great precision; e.g., McCarthy of the internal politics of a small United States college in "The Groves of Academe" or Snow's "The Masters" with its model of a Cambridge college. (The ambiguity--and richness--of verbal models is evident when one considers the totally different overtones of the one word "college.") He is in line with Maynard Smith's use of the word when he writes: "Good mathematical models don't 'predict' in the colloquial sense of the word. But they can broaden our understanding of the potential consequences of our decisions..." In a very different context from the Venice seminar of economists, sociologists and futurists to whom Toffler spoke, Oster and Guckenheimer have reported [128, p. 328] "that many ecologists seriously question whether mathematics can play any useful role in biology. Some claim that there has not been a single fundamental advance in biology attributable to mathematical theory. Where complex systems are concerned, they assert that the appropriate language is English, not mathematical." These are reservations worth bearing in mind for mathematical modelling is not without a certain danger of narcissism (cf. [224]).

The use of mechanical analogies, verbally described, is not uncommon in mathematical modelling. Rheologists will, for example, often talk about springs and dashpots even when the equations they obtain could have been derived directly from more abstract hypotheses. When they draw further pictures of the coiling and uncoiling of polymer chains they may indeed be making mechanical hypotheses. But analogies also have their place. How much of a place is to be given to them is partly a matter of taste, and there is a famous passage in Duhem [56] contrasting like French and English in their approach to the physics of his day. "The employment of similar mechanical models...is a regular feature of the English treatises on physics. Here is a book intended to expound the modern theories of electricity...in it there are nothing but strings which move around pulleys, which roll around drums, which go through pearl beads, which carry weights; and tubes which pump water while others swell and contract; toothed wheels which are geared to one another and engage hooks. We thought we were entering the tranquil and neatly ordered abode of reason, but we find ourselves in a factory."

2.2 Finite models.

The theory of graphs has found many natural applications in the physical and social sciences. A graph is a collection of vertices, V, linked in some way by elements of a collection of edges, E. If u, $v \varepsilon V$ are vertices (the terms node and point are also used), then $uv \varepsilon E$ denotes the edge connecting u to v. If this has a sense of direction from u to v, the graph is a directed graph or digraph. E is thus a subset of VxV. If vv is allowed as a proper element of E, then the graph is called a loop graph. If more than one edge may connect two vertices, the term multiple graph or multigraph is often used. The problems of connectivity and decomposition, or characterization and the topography of graphs arise immediately and there is a rich theory on which to draw. The most obvious applications are to networks of all sorts, electrical, mechanical, transportation, job assignment and scheduling, industrial interdependence and the planning of experiments. Perhaps the most famous "application" is the four-colour problem, the many, unsuccessful attempts to solve which have contributed greatly to the advancement of graph theory. But the structure of many different areas, social, physical and intellectual, can be illuminated by its methods; see, for example [85]. It has even been applied to the structure of Mozart's "Cosi fan Tutte" [83]. It has a vast literature of which only a sample can be mentioned: [24,27,28,67,84,85,140].

The theory of games is another finite model that has found wide application and indeed was developed by von Neumann and Morgenstern in the context of economics. In its simplest form two players P and Q each select one of their set of options $(p_1,\ldots p_m)$ for P and $(q_1,\ldots q_m)$ for Q, the payoff from P to Q if the choices are p_i and q_j being an amount a_{ij}. The question is whether there is an optimal way for each to play. A pure strategy for p is the choice of one p_i, a mixed strategy, σ, is the choice of a set of numbers s_i, $i=1,\ldots m$, $\Sigma s_i=1$, which can be regarded as the proportion of times in a long run that p_i will be chosen. A similar definition applies for t_j, a mixed strategy τ for Q. The expected payoff $p(\sigma,\tau) = \Sigma\Sigma s_i a_{ij} t_j$ and the game has a value if there exist two strategies $\bar{\sigma}$ and $\bar{\tau}$ such that for all σ and τ

$$p(\bar{\sigma},\tau) \geq p(\bar{\sigma},\bar{\tau}) \geq p(\sigma,\bar{\tau}).$$

The fundamental theorem shows that a two-person game has a value under the condition that it is zero-sum, i.e., the payoff to Q of an amount a is the same as a payoff to P of -a. There are many natural applications in the social and physical sciences. Some problems in control theory can be regarded as games against Nature. There are also many extensions, as to non-zero-sum and multiperson games and to differential games in which the action and condition of the players develop continuously in time. See, for example, [69,99,121,149,150,191].

Game theory is intimately connected with linear programming, the problem of determining the set of non-negative x_j, $j=1,2,\ldots N$, that satisfy

$$\sum_{j=1}^{N} a_{ij}x_j \leq b_i, \quad i=1,2,\ldots,M$$

and maximize

$$Z = \sum_{j=1}^{M} c_j x_j.$$

To which there is a dual problem, that of minimizing

$$W = \sum_{i=1}^{M} b_i y_i$$

subject to

$$\sum_{i=1}^{M} y_i a_{ij} \geq c_j, \quad j=1,2,\ldots N.$$

Many systems have been modelled in this form ranging from huge input-output models of the economy to modest problems of blending. Dantzig, the principle architect of the subject, has given a splendid exposition of it [52] and its extensions in a book from which one can learn much about modelling in general.

Finite automata have been used variously as models, not only in computer science, the house in which they were born, but also in control theory, linguistics, psychology, and biology. See [8,9,62,78,168,180]. Basically, the automaton is the computer reduced to its simplest elements, an input/output tape on which symbols from a finite alphabet are read or written and a set of internal states. The computation is done by a set of instructions that modify the internal state and either move the tape or modify the symbol under the head. If the alphabet is $A = \{a_j;\ j=1,\ldots M\}$ and the states are a set $S = \{S_k;\ k=1,2,\ldots N\}$ the instruction can be of three forms: $(a_j, S_k) \rightarrow (a_p, S_q)$ says that when the machine is in state S_k and reads a_j it replaces the symbol by a_p and changes its state to S_q; $(a_j, S_k) \rightarrow (1, S_q)$ changes the state to S_q but leaves a_j unchanged merely moving the tape one place to the left; $(a_j, S_k) \rightarrow (r, S_q)$ does the same except that movement is to the right. The machine is deterministic if (a_j, S_k) has a unique consequence. A computation is a series of such steps which either terminates or goes into a repetitive sequence. Usually there is a distinguished initial state and the input tape, which is finite, starts in the left-most position. When it stops, the state of the tape may be regarded as the output. The Turing machine, as this automaton is called, thus gives a mapping from the input tape to the output tape. It can be generalized to have several reading and writing heads, but it is a matter of convenience rather than necessity for Church's thesis is that all devices that formalize the notion of computability are equivalent. The finite automaton is a Turing machine that only reads and only moves the tape in one direction, say to the left; its instructions can therefore be written $(a_j, S_k) \rightarrow S_q$ since the l can be taken for granted. It is not hard to see how such a device has possibilities for representing the learning process or the formal aspects of language and these and other developments are to be found in the references given.

2.3 Fuzzy subsets.

An important class of mathematical models was introduced by Zadeh [195] in 1965 when he defined a fuzzy set or subset. The usual definition of a subset A of U can be formalized in terms of the characteristic function of the subset $\chi(x)$ such that for any $x \in U$, $\chi(x)=1$ if $x \in A$ and $\chi(x)=0$ if x does not belong to A. Such definiteness is all very well in its place but there are clearly many situations whose intrinsic ambiguity and vagueness is ill-served by such a black-or-white attitude. The concept of fuzzy subset replaces the characteristic function $\chi(x)$ with values in the set $\{0,1\}$ by a membership function $\mu(x)$ with values in a membership set M. Usually M is a totally ordered set (very often the closed interval (0,1) is taken) and $\mu(x)$ is the degree of membership of x in A. For example, if U is the real line and M=[0,1], the fuzzy subset of "small numbers" might be defined with $\mu(x)=(1+|x|)^{-1}$ or of "really small numbers" by $\mu(x)=(1+|x|)^{-50}$.

The usual operation with sets can be defined suitable for fuzzy subsets. For example, if A and B are fuzzy subsets of U with the same membership set, A is included in B is $\mu_A(x) \leq \mu_B(x)$ for all $x \in U$. The union of two fuzzy subsets has a membership function $\mu(x) = \mathrm{Max}(\mu_Z(x), \mu_B(x))$. For example, if M and U are both [0,1], these ideas can be shown graphically.

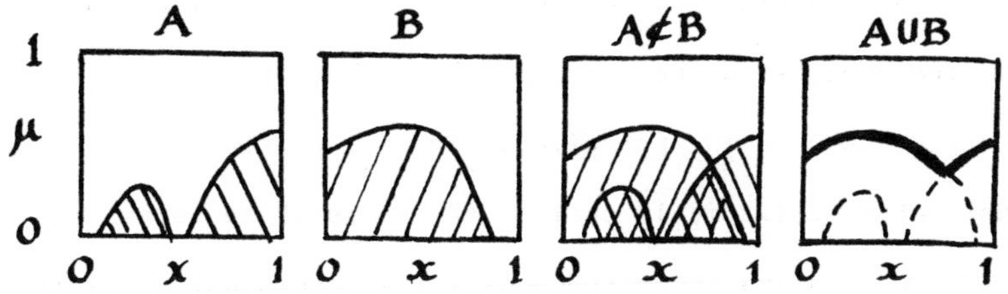

In the same spirit graphs, relations and equivalence can be generalized to fuzzy graphs, fuzzy relations and similitude. Binary operations on two fuzzy subsets can be defined leading to monoids and groupoids. For example, if $U=R_+$ and I_n is the fuzzy subset with membership function $\mu_n(x) = \lambda^n x^{n-1} e^{-\lambda x}/(n-1)!$ then we can define the composition of two of these subsets $I_m * I_n$ as the fuzzy set with membership function

$$\int_o^x \mu_m(x-t)\ \mu_n(t)\ dt$$

Then $I_m * I_n = I_{m+n}$ and, if we add I_o with $\mu_o(x) = \delta(x)$, we have a monoid $I_o, I_1 \ldots$ of fuzzy subsets which is isomorphic to the natural numbers. The I_n are called the exponential fuzzy integers.

But the generalization have gone far beyond this and there are fuzzy categories, topological spaces, logics, algorithms, automata, languages, and environments. Some of the most important applications concern decision-making in a fuzzy environment; see, for example, Bellman and Zadeh [26]. The book of Kaufmann [104] is a lucid introduction; Negoita and Ralescu [134] are very good but poorly translated; see also Bellman and Giertz [25] and Zadeh, Fu, Tanaka and Shimura [192].

2.4 Statistical models.

In his analysis of a system S, Bury [39] lumps together the process and its measuring devices as a black-box out of which comes perceived data. The data generated by the process can be corrputed by either systematic or random error in the process of being perceived. The aim of statistical analysis is thus to construct a statistical model on the basis of the available output and hence form conclusions about the underlying phenomena. This is to view the measurements as realization of a random variable X. The statistical model of the process is thus the probability density function (sometimes called the probability mass function when it associates non-zero probabilities with points) or its integral, the cumulative distribution function, for the random variable. The statistical tools needed to make the desired inferences are an understanding of sampling theory and order statistics and the various qualities of inference--consistency, bias, minimum variance, efficiency, etc.--and the several estimation and confidence tests. There is a range of distributions (Gaussian or normal, log-normal, gamma, beta, binomial, Poisson, Weibull), each with its own virtues and properties.

Of the vast array of books on probability and statistics it is almost impertinent to single out one or two. Kendall and Stuart's magisterial volumes may still be given pride of place [106], but Feller [64], Lindley [117] and Parzen [143] might be mentioned. There is, of course, a huge literature on the philosophical issues in "the matter of chance" [132].

2.5 Difference and differential equations.

Differential equations will so dominate the rest of these notes that it would be somewhat gratuitous to do more than mention them here. They are by far the commonest type of model in the physical sciences, ordinary differential equations playing the same role for lumped systems as partials do for distributed. Whole tracts of control theory, for example, can be cast in the form

$$\dot{x} = f(x,u,t), \quad y = g(x,u,t), \tag{2.1}$$

where $x = x(t)$ is a vector of state variables, y a vector of observations and u a vector of control variables. Feedback control seeks a function $u = h(y)$ which will attain certain goals; optimal control theory is concerned with finding the control u which will maximize or minimize some functional of the path or function of the final state. There are key questions of observability and controllability in such models which lie at the root of the question of whether optimal or feedback controllers exist. Naturally these questions are answered most completely for linear systems. Similar questions pertain to partial differential equations where they are of course much more difficult to answer.

Difference equations are appropriate when the dependent variable is discrete. An example has already been given of the logistical equation $x_{n+1} = \lambda x_n(1-x_n)$. Sampled continuous systems are also an avenue to different equations. Their theory parallels that of differential equations in many ways though the example just given does suggest that their behavior can become bizarre much earlier. The difference-differential equation

$$\dot{x}(t) = f(x(t), x(t-1)) \tag{2.2}$$

is an example of a functional differential equation. Whereas eqn. (2.1) requires initial values $x(0)$ to be specified if a particular solution is to be determined, eqn. (2.2) requires all values $x(t)$ over a unit initial interval, say $-1 \le t \le 0$. In that the differential equation has to be discretized when its solution is to be computed on a digital machine it is clear that the difference equation is more important than one would first suppose.

The solution of $\dot{x} = f(x,t)$ also satisfies

$$x(t) = x(0) + \int_0^t f(x(t'),t')dt' \tag{2.3}$$

which is an integral equation. It can be regarded as a nonlinear integral equation and if X is a suitable defined class of functions the right hand side of eqn. 2.3 is a nonlinear transformation from X into X. If this transformation has a fixed point it provides a solution of the integral equation. Integral equations and indeed integro-differential equations are used as models, though we shall not have occasion to refer to them.

2.6 Stochastic models.

The integral formulation (2.3) of the differential equation is often used as a starting point for the study of differential equations with random elements. Thus if α is a random variable, i.e. known only through its probability distribution function $F(a) = \Pr\{\alpha<a\}$ or its density $f(a)da = \Pr\{\alpha\varepsilon(a, a+da)\}$, a generalization of eqn. (2.1) is

$$x(t;\alpha) = f(x(t;\alpha),t;\alpha) + g(x(t;\alpha), t)w(t;\alpha) \tag{2.4}$$

where w is a random process often taken to be white noise. The initial state is also random $x(0;\alpha) = x_o(\alpha)$ but we have a formal analogue of eqn. (2.3) in

$$x(t;\alpha) = x_o(\alpha) + \int_0^t f(x(t';\alpha), t';\alpha)dt + \int_0^t g(x(t';\alpha),t')dW(t';\alpha) \tag{2.5}$$

where W is the Wiener process from which w is derived. Unfortunately the second integral almost surely does not exist in the ordinary Riemann-Stieltjes sense and it requires special interpretation (see [100],[126] etc.). Nevertheless it is clear that in such stochastic differential equations we have a class of model of the utmost importance. Nor can we fail to expect some striking new results. Take for example the very simplest of cases, the equation $\dot{x} + \alpha x = 0$, $x_o=1$ where α is uniformly distributed over the interval $(\beta-\gamma, \beta+\gamma)$. The expected value of α is β and with this expected value for α the solution of the equation is $x(t) = \exp-\beta t$. But $x(t;\alpha) = \exp-\alpha t$ and the expected value of this is $E(x(t;\alpha)) = (\exp-\beta t)[(\sinh\gamma t)/\gamma t]$ which behaves quite differently as time goes on. Seldom can the equation be solved explicitly and one has recourse to moments or equations for the transition probability. Thus if $F(x,t|x_o,t_o)=\Pr\{x(t;\alpha)<x|x(t_o;\alpha)=x_o\}$ and $f = \partial F/\partial x$ and $\dot{x} = m(x,t) + \sigma(x,t)w(t,\alpha)$ then F satisfies a Kolmogoroff or backwards diffusion equation and f the Fokker-Planck equation

$$\frac{\partial f}{\partial t} = \frac{1}{2}\frac{\partial^2}{\partial x^2}(\sigma^2 f) - \frac{\partial}{\partial x}(mf)$$

with a boundary condition from the probability density function of the initial state.

The origins of the theory of stochastic processes lie in the study of Brownian motion and other random walk problems, though Markov, one of the key figures in the development, had interests in linguistic problems also. We have mentioned some elementary stochastic models in Appendices A and B. A key notion of wide applicability is that though only known probabilistically the state at time $t+1$ only depends on the state at time t. This property, often called the Markov property, means that the transition probability between state i at time t and stage j at time $t+1$, $p_{ij}(t)$, is all that need be known. If it is independent of t the process is stationary. The theory develops from random walk problems to Markov chains with discrete time and a finite number of states, to processes with discrete states in continuous time and so to those with continuous state space and continuous time, i.e. stochastic differential equations. Non-Markovian processes can also be considered though with more difficulty. Time series, the outputs of stochastic processes, are also studied for their own sake and prediction and filtering theory plays a key role in many applications. It is easy to see that stochastic models are appropriate to a wide range of situations from learning theory or the mobility of the work force to gunnery and ecology. An excellent bibliography up to 1959 has been edited by Wold [192] and there is an extraordinary range of books on the subject of which a few are: [22, 44,48,100,103,126,139,143,154,166,193,227]. For a graphic presentation of three stochastic processes the introduction to [192] makes interesting reading.

3 How to formulate a model

"'You may seek it with thimbles--and seek it with care;

You may hunt it with forks and hope;

You may threaten its life with a railway share;

You may charm it with smiles and soap--'"

C. L. Dodgson. The Hunting of the Snark.

Fit. 3, St. 8.

Comparatively little needs to be said on this score now that we have reviewed the types of model that are available for the formulation is nothing more than a rational accounting for the various factors that enter the picture in accordance with the hypotheses that have been laid down.

3.1 Laws and conservation principles.

The formulation of the equations of a model is usually a matter of expressing the physical laws or conservation principles in appropriate symbols. This can often be written down as a prescription as, for example, in particle dynamics [107] where, if m is the mass of the particle and $\underset{\sim}{r}$ its position with respect to an inertial frame, $m\ddot{\underset{\sim}{r}}$ is calculated and set equal to $\underset{\sim}{F}$, the resultant of all forces acting on the particle. This is a second order equation and, this being recognized, it will clearly be necessary to specify the initial position, $\underset{\sim}{r}_o$, and velocity, $\dot{\underset{\sim}{r}}_o$, before the model is complete. For a single rigid body whose center of mass is at $\underset{\sim}{s}$ in an inertial frame, $m\ddot{\underset{\sim}{s}}$ is equated to the resultant of the forces and $\dot{\underset{\sim}{H}}_Q$, the rate of change of the vector of moments of momentum, to the moment of these forces about Q, provided that Q is fixed or the center of mass, $\underset{\sim}{s}$.

Except in relativistic contexts when the interconvertibility of mass and energy is at issue, the conservation principles invoked for physical problems are usually those of mass, momentum or energy. We can also do number counts, as, for example, in a population model which may be used to illustrate the basic kind of balance that is involved. Let $n(a,t)da$ be the number of individuals in the age bracket $(a,a+da)$, then we can compute the change in this number during the time interval $(t,t+dt)$. By definition of n this is $\{n(a,t+dt)-n(a,t)\}da$ and by a balance over the age bracket this is the number of individuals who 'age' into the bracket, i.e., $n(a-dt,t)dt$, minus the number who age out of it, i.e., $n(a,t)dt$, minus those that die $\Theta(a,t)dadt$. Thus dividing by dadt

$$\frac{n(a,t+dt)-n(a,t)}{dt} = \frac{n(a-dt,t)-n(a,t)}{da} - \Theta(a,t).$$

We then recognize that age and time run simultaneously so that $da=dt$. Letting the common value of this increment go to zero we have

$$\frac{\partial n}{\partial t} + \frac{\partial n}{\partial a} + \Theta = 0. \tag{3.1}$$

A more sophisticated version of this basic balance starts by recognizing the simultaneous flow of age and time. This implies that $n(a,t)$ is the flux of numbers that age across a at any time t. Thus, for any interval of ages $b<a<c$ (not necessarily small), $\int_b^c n(a,t)da$ is the total number in that interval, $n(b,t)-n(c,t)= -\int_b^c(\partial n/\partial a)da$ is the net flux into the interval and $\int_b^c \Theta(a,t)da$ is the total loss by death. Thus

$$\frac{d}{dt}\int_b^c n(a,t)da = \int_b^c \frac{\partial n}{\partial t}\, da = -\int_b^c \frac{\partial n}{\partial a}\, da - \int_b^c \Theta\, da$$

or

$$\int_b^c \left[\frac{\partial n}{\partial t} + \frac{\partial n}{\partial a} + \Theta\right] da = 0.$$

But if the integrand is continuous it must vanish everywhere, for, suppose it were positive at a_o, $b \leq a_o \leq c$, then it would be positive in some interval about a_o. But then b and c could both be taken in this interval and the integral could not be zero. The recognition of n(a,t) as a flux across the age a makes it easy to write a boundary condition since n(0,t) is the rate of total births. Thus, if $\gamma(a,t)da$ is the number of births per unit time to individuals in the age bracket (a,a+da) at time t,

$$n(0,t) = \int_0^\infty \gamma(a,t)n(a,t)da. \tag{3.2}$$

This process can be stated rather generally as follows. In a discrete element we can let, F be the net flux of the entity into the element, G its rate of generation there and H the total amount of it which is present. Then F, G and H are functions of time and satisfy

$$F + G = \frac{dH}{dt} \tag{3.3}$$

If we are dealing with a continuum then these quantities must be defined as densities. Thus we let the vector $\underset{\sim}{f}$ denote a flux which is defined such that the flux across an element of area dS in the direction of its normal n is $\underset{\sim}{f} \cdot \underset{\sim}{n}\, dS$. Similarly the generation must be defined as a rate per unit volume, so that in a volume element it is gdV, and H becomes a concentration h. Then if Ω is an arbitrary, simply connected region of the continuum with a piece-wise smooth surface $\partial\Omega$ whose outward normal is denoted by $\underset{\sim}{n}$, we have

$$-\int\int_{\partial\Omega} \underset{\sim}{f} \cdot \underset{\sim}{n}\, dS + \int\int\int_{\Omega} g dV = \frac{\partial}{\partial t} \int\int\int_{\Omega} h dV$$

In this equation we use the fact that Ω is fixed to interchange the order of integration and differentiation and use Green's theorem on the surface integral. Then all terms can be brought to one side of the equation and we have

$$\iiint_{\Omega}\left[\frac{\partial h}{\partial t} + \nabla \cdot \underset{\sim}{f} - g\right] dV = 0.$$

We must now make the hypothesis that $\underset{\sim}{f}$, g and h are sufficiently continuous that the integrand is continuous and then, since the region Ω is completely arbitrary,

$$-\nabla \cdot \underset{\sim}{f} + g = \frac{\partial h}{\partial t}. \tag{3.4}$$

If the volume is a material volume $\Omega(t)$ moving in a continuum where the velocity field is $\underset{\sim}{v} = \underset{\sim}{v}(\underset{\sim}{x},t)$, then we need Reynolds' theorem for the interchange of differentiation with respect to time and integration. This is

$$\frac{d}{dt}\iiint_{\Omega} h dV = \iiint_{\Omega}\left[\frac{\partial h}{\partial t} + v\nabla\cdot h\right] dV.$$

The fact that the flux through a surface element can always be expressed as $\underset{\sim}{f}\cdot\underset{\sim}{n}\, dS$ is the conclusion of an interesting type of argument that is sometimes useful in other contexts. The figure shows a particular form of element, namely a tetrahedron of volume dV and with three sides perpendicular to the axes On_1, On_2, On_3 and having dS as the area of its slanting face. Then by definition of the direction cosine the face perpendicular to On_1 has area $n_1 dS$. Let f_i be the flux in the direction On_1 and f the flux over the slant face. Then a balance can be struck over the tetrahedron for

$$F = (f_1n_1 + f_2n_2 + f_3n_3 - f)dS,\ G = gdV,\ H = hdV$$

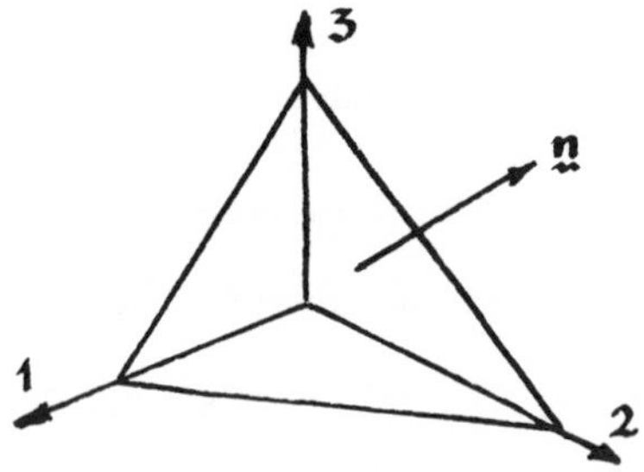

and the equation $F + G = \dot{H}$. But if the volume is allowed to shrink in size, whilst keeping its proportions, dS will decrease as the square of the size but dV as the cube. It follows that G and H become negligible in comparison with F and hence in the limit F = 0. Thus

$$f = f_1 n_1 + f_2 n_2 + f_3 n_3 = \underset{\sim}{f} \cdot \underset{\sim}{n} \tag{3.5}$$

if f is the vector with components (f_1, f_2, f_3).

A similar argument is used in the formulation of boundary conditions if the element over which the balance is made is an element of surface extended by a distance dh on either size. Then letting $dh \to 0$ first reduces

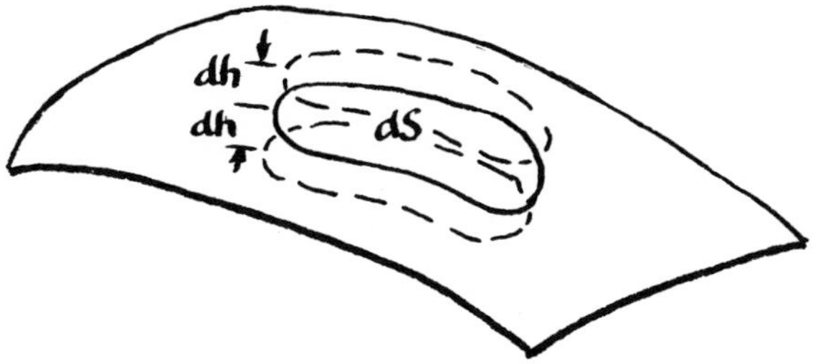

the volume to zero. It is another way of saying that in three dimensions a surface has no volume and hence no capacity for a quantity defined per unit volume. Thus in the boundary conditions (C4) and (C5) there is no term with a time derivative since the natural capacity of the surface is zero. If however a quantity is defined 'per unit area' and is therefore a surface concentration it may well show up in a boundary condition.

The same principles apply to a moving discontinuity as can best be seen in one dimension for a kinematic wave moving in the x-direction

$$\frac{\partial f}{\partial x} + \frac{\partial h}{\partial t} = g$$

If a discontinuity moves with velocity u having a flux f_- and concentration h_- to the left and f_+, h_+ to the right then in a time dt the net flux $(f_- - f_+)dt$ provides the amount that builds up as the front advances, which in (t,t+dt) is $(h_- - h_+)udt$. Thus the speed of the discontinuity is

$$u = \frac{f_+ - f_-}{h_+ - h_-}$$

We observe that there is no term in g here for there can be no contribution from the discontinuity itself which is a set of measure zero. Put another way, we could say that the amount generated would be $gdxdt = gu(dt)^2$ which becomes vanishingly small in comparison with the other terms as $dt \to 0$.

3.2 Constitutive relations.

In formulating a general conservation relation we left the relationship between flux and concentration undefined. This is desirable since the physical laws (of conservation of matter, etc.) are applicable to a variety of substances of different constitutions. It is the duty of the constitutive relation to provide the connection between f, g, and h or F, G and H. Thus for example in the stirred tank model of Appendix C when the heat equation, (C3),

$$\rho_w c_{pw} \frac{\partial T_w}{\partial t} = k_w \nabla^2 T_w$$

is written for the wall, we assert that

$$\underset{\sim}{f} = -k_w \nabla T_w, \; g=0, \; h=\rho_w c_{pw} T_w$$

for the entity 'energy' or 'heat'. Thus the constitution of the wall is that it conducts heat according to Fourier's law, $f \propto -\nabla T$, and does not of itself generate heat, $g = 0$. Similarly the species balance on the whole reactor, eqn. C1,

$$V \frac{dc_j}{dt} = q_j c_{jf} - qc_j + \alpha_j Vr(c_1, \ldots c_s, T)$$

is of our general form with

$$F = q_j c_{jf} - qc_j, \quad G = \alpha_j Vr(c_1, \ldots c_s, T), \quad H = Vc_j.$$

The expressions for F and H follow from the definitions but that for G asserts two things about the constitution of the system; first, that the reaction rate is a function of all the concentrations and the temperature; second, that there is but one reaction in which the stoichiometric coefficient of A_j is α_j. This is further specialized by the constitutive relation of the first order irreversible reaction in Σ_6 where $\alpha = -1$, $r = k(T)c$.

Under certain circumstances it might be well to distinguish even the generally applied constitutive relations from the physical laws in the system of hypotheses. This has not been done in the appendices where Fourier's, Fick's law or Newton's law of cooling have been lumped with the basic laws in an underlying hypothesis, H_0, but the more specific constitutive relations have been made explicit (e.g. H_2 and H_{10} for the reaction rate in Appendix C).

Constitutive relations may come in alternative forms as when Fick's law expresses the diffusive flux in terms of concentration gradients whereas the Maxwell relations for multicomponent diffusion express these gradients in terms of the fluxes. In many cases the forms can be converted into one another and one should keep an unprejudiced mind in case a conversion is desirable, but in some cases there are natural choices. Thus in a problem of diffusion and first-order reaction

$$D\nabla^2 c = kc \text{ in } \Omega$$

$$D\underset{\sim}{n}\cdot\nabla c = h(c_f - c) \text{ on } \partial\Omega$$

one could formulate the problem in terms of a flux

$$\underset{\sim}{j} = -D\nabla c$$

since then

$$c = -\frac{1}{k}\nabla\cdot\underset{\sim}{j}$$

and

$$D\nabla^2\cdot\underset{\sim}{j} = k\underset{\sim}{j} \text{ in } \Omega$$

with

$$-\underset{\sim}{n}\cdot\underset{\sim}{j} = h + (h/k)\nabla\cdot\underset{\sim}{j} \text{ on } \partial\Omega.$$

Though the second set of equations has three components and is clearly not the set to use to solve the problem yet the complementary formulations have their roles in the variational properties of the solutions since the solution c minimizes the functional

$$\iiint_{\Omega}\left[D(\nabla c)^2 + kc^2\right] dV + \iint_{\partial\Omega} h\left[(c_f - c)^2\right]$$

whilst $\underset{\sim}{j}$ maximizes

$$2\iint_{\partial\Omega}(\underset{\sim}{n}\cdot\underset{\sim}{j})dS - \iiint_{\Omega}\left[\frac{1}{D}\underset{\sim}{j}^2 + \frac{1}{k}(\nabla\cdot\underset{\sim}{j})^2\right] dV - \iint_{\partial\Omega}\left[\frac{1}{h}(\underset{\sim}{n}\cdot\underset{\sim}{j})^2\right] dS$$

Many examples of this duality are to be found in Arthurs [19].

Powerful general principles can often be brought to bear on constitutive relations to show what general form they must have. Thus Serrin [160] shows that the stress tensor, $\underset{\sim}{T}$, of a Stokesian fluid must be related to the rate of strain tensor, $\underset{\sim}{D}$, in the form

$$\underset{\sim}{T} = \alpha\underset{\sim}{I} + \beta\underset{\sim}{D} + \gamma\underset{\sim}{D}^2$$

where α, β and γ may be functions of the three invariants of $\underset{\sim}{D}$. Examples of this kind of reasoning abound in rational mechanics, as, for example, in Truesdell and Toupin's materpiece [179].

3.3 Discrete and continuous models.

At this point it is useful to discuss the alternative formulation of discrete and continuous models of physical problems. Each has its advantages and disadvantages but even here the distinctions are not absolute nor clean-cut, either in substance or in method, and may manifest themselves at one time as ordinary vs. partial differential equations, at another in linear algebraic vs. integral equations. Moreover when it comes to computation on a digital computer, the continuous necessarily becomes discrete. The terms 'lumped parameter' and 'distributed parameter' systems seem misguided for it is variables not parameters that are lumped (discrete) or distributed (continuous).

The word 'lumping', in spite of its ungainly overtones, is useful in describing the process by which a number of things are put together in one. This may result in replacing a continuous system by a discrete one. An example of this is the network thermodynamics of Oster, Perelson and Katchalsky [141], where problem of flow and transport in biological systems are treated by the ideas of electrical network theory. This converts parabolic equations into ordinary differential equations and elliptic into algebraic. An important discussion of lumping in the context of monomolecular reactions has been made by Wei and Kuo [190]. This reduces a large system of species into a continuum of species and in this sense lumps them together. Lumping of this sort does replace a large number of equations by a single equation, but this is often an integro-differential equation rather than an ordinary differential equation. In fact the sense of the word has been turned around and what is being done is the distribution of a large number of discrete variables into a continuous variable--vigorously stirring out the lumps. These two processes are worth considering further as they illustrate some of the subtleties of the relations between continuous and discrete models. Consider first the loss of heat of a wall to its environment--a thermal analogue of Oster's case of diffusion through a membrane. The figure below shows the physical picture with $T(x,t)$,

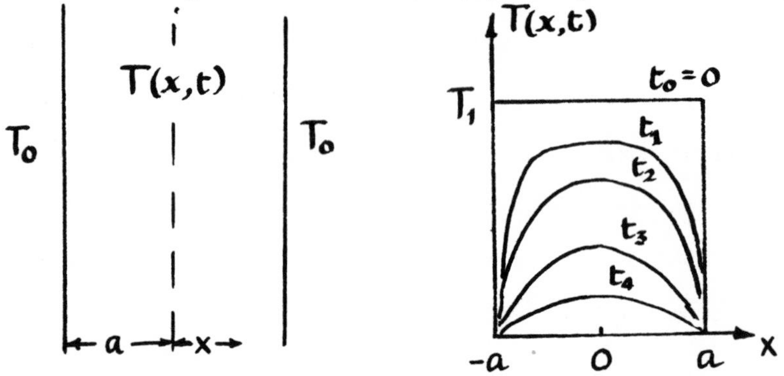

$-a \le x \le a$, the temperature at any point of the wall. Thus (incorporating the symmetry) the continuous model be

$$\begin{aligned} \rho c_p \frac{\partial T}{\partial t} &= k \frac{\partial^2 T}{\partial x^2}, \quad 0 < x < a, \\ \frac{\partial T}{\partial x} &= 0, \quad x = 0, \\ k \frac{\partial T}{\partial x} &= h(T_o - T), \quad x = a, \\ T(x,o) &= T_1. \end{aligned} \tag{3.6}$$

The kind of solution that we expect for these equations is sketched in the right part of this figure. Let us make the equations dimensionless by writing

$$\xi = x/a, \quad \tau = kt/\rho c_p a^2, \quad u = (T-T_o)/(T_1-T_o), \quad \mu = ha/k. \tag{3.7}$$

Thus

$$u_\tau = u_{\xi\xi}, \quad 0 < \xi < 1,$$

$$u_\xi = 0, \qquad \xi = 0,$$

$$u_\xi + \mu u = 0, \xi = 1, \tag{3.8}$$

$$u(\xi,0) = 1.$$

The complete solution of this equation is elementary enough, being

$$u(\xi,\tau) = \sum_1^\infty \frac{2\mu}{\lambda_n^2(\sec\lambda_n + \sin\lambda_n)} e^{-\lambda_n^2\tau} \cos\lambda_n\xi \tag{3.9}$$

where

$$\lambda_n \tan\lambda_n = \mu, \; n = 1,2,... \tag{3.10}$$

In particular the average temperature is

$$\bar{u}(\tau) = \sum_1^\infty \frac{2\mu^2}{\lambda_n^2(\lambda_n^2 + \mu\lambda_n + \mu^2)} e^{-\lambda_n^2\tau}. \tag{3.11}$$

The sequence of eigenvalues λ_1, λ_2 increases rapidly; for example with large μ, $\lambda_n = (2n-1)\pi/2$. Thus all the terms after the first quickly became negligible and

$$\frac{d}{d\tau}\bar{u}(\tau) \sim -\lambda_1^2\,\bar{u}(\tau). \tag{3.12}$$

This makes it look like a first order system with a time constant of λ_1^{-2}. In terms of real time this is

$$a^2/\lambda_1^2 D, \tag{3.13}$$

or $4a^2/\pi^2 D$ as $\mu\to\infty$.

The network analogue consists in lumping the resistances and examining the driving potentials. A wall of unit area, thickness 2a and conductivity k will conduct heat at a rate $Q = k\Delta T/2a$ and might therefore be regarded as having a resistance of $\Delta T/Q = 2a/k$. Let this resistance which is in fact distributed over the thickness, be divided into two and lumped at the surfaces of the slab as shown in the figure. The temperature within the slab sill also have to be lumped and since the resistance has been separated to the walls the natural lumping is $\bar{T}$, the mean temperature. The flow of

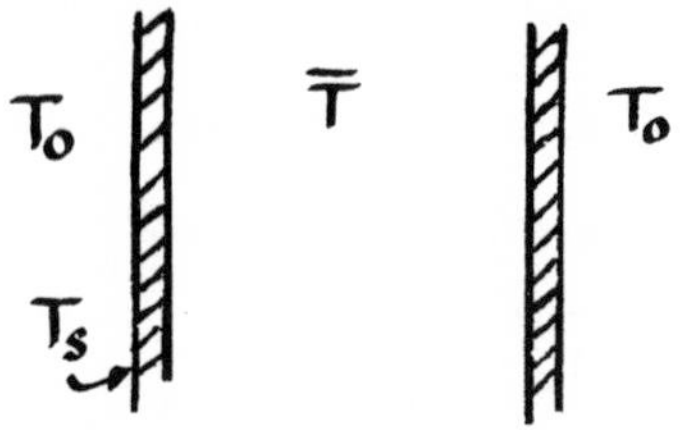

heat out of the slab at each face will therefore be $(k/a)(\bar{T}-T_s)$ where T_s is the temperature at the outside surface of the slab. This just matches the flow through the external film $h(T_s-T_o)$ so we have

$$q = \frac{k}{a}(\bar{T}-T_s) = h(T_s-T_o) = \frac{k}{a}\left(1+\frac{1}{\mu}\right)^{-1}(\bar{T}-T_o).$$

Now the thermal capacity of the slab is clearly $2a\rho c_p$, so

$$2a\rho c_p \frac{d\bar{T}}{dt} = 2\frac{k}{a}\left(1+\frac{1}{\mu}\right)^{-1}(T_o-\bar{T}) \tag{3.14}$$

or

$$\frac{d\bar{u}}{d\tau} = -\left(1+\frac{1}{\mu}\right)^{-1}\bar{u} \tag{3.15}$$

This gives a dimensionless time constant of $(1 + 1/\mu)$ or in dimensional terms

$$a^2(1+\mu)/D\mu, \tag{3.16}$$

and in the limit $\mu \to \infty$,

$$a^2/D. \tag{3.17}$$

Comparing (3.13) and (3.17) we see that we are off by a factor of two and a half. This suggests that agreement could be improved by assuming that the "dynamic" capacity of the wall is only two fifths of its "static" capacity $2a\rho c_p$. For the more general case of external resistance the ratio of the time constants of the lumped and distributed systems is

$$t_L/t_D = \mu/(1 + \mu)\lambda_1^2$$

$$= \frac{\tan\lambda_1}{\lambda_1(1+\lambda_1\tan\lambda_1)}$$

As μ decreases from infinity to zero, λ_1 decreases from $\pi/2$ to 0 and the ratio increases from $(4/\pi^2)$ to 1. It is not surprising that the two time constants should agree in the limit $\mu \to 0$, or $k/a \ggg h$, for if the resistance of the outside films dominates so completely then the system is truly lumped.

This comparison illustrates some of the difficulties in going from the discrete to the continuous. A certain amount of accuracy can be recovered by the use of a pseudo-capacity, much as a virtual mass can be used in other cases, but it is not altogether satisfactory to have the wall capacity depend on the heat transfer coefficient. An analogous method of reducing the partial differential equations of a catalyst particle to ordinary differential equations was used by Hlavácek et al. [95]. Reference to this is given in [12] and [14] where the one-point collocation method of Villadsen and Stewart [188] is also described. (See also Sec. 4.5.2 below and [206]).

To illustrate the kind of "lumping" that is really distribution consider Luss and Hutchinson's [123] treatment of many parallel first order reactions. In many situations it is not possible to describe a mixture of chemical species that boil between, say, 350° and 500° and this might be taken as a largish lump. On the other hand if we talk about the number of moles $n(T)dT$ that boil in the range $(T, T + dt)$ we have really made a continuum, i.e. an infinity of species, out of a system that is necessarily discrete. In the case of species which can all undergo a reaction $A_i \to B_i$ with rate constant k_i we may devise a continuum and talk about the "species" $A(k)dk$ as all that reacts with rate constant in the range $(k,k + dk)$. If $c(t,k)dk$ is the concentration of this material at time t and if the reactions are all parallel first order

$$c(t,k)\, dk = c(o,k)\, dk\, e^{-kt} \tag{3.18}$$

Now in many cases we may only be interested in the total amount $C(t) = \int_o^\infty c(t,k)dk$ and we see that

$$C(t) = \int_o^\infty e^{-kt} c(o,k)dk \tag{3.19}$$

is the Laplace transform of the initial distribution with time, for a change, playing the role of the transform variable.

It is interesting to enquire if there is an apparent rate law

$$C = \frac{dC}{dt} = -f(C) \tag{3.20}$$

but although $\dot{C} = -\int_o^\infty ke^{-kt}c(o,k)dk$ it is seldom possible to invert (3.19) and so eliminate t. A notable exception is

$$c(o,k) = C(o)k^\alpha e^{-\beta k} \frac{\beta^{\alpha+1}}{\Gamma(\alpha+1)}, \quad \alpha,\ \beta > 0. \tag{3.21}$$

Let us make c dimensionless by dividing by $C(o)$, i.e.

$$u(t,k) = C(t,k)/\int_o^\infty c(o,k)dk \tag{3.22}$$

so that for this distribution

$$u(o,k) = \beta^{\alpha+1}k^\alpha e^{-\beta k}/\Gamma\ (\alpha+1) \tag{3.23}$$

Then by eqn. (3.19)

$$U(t) = \frac{C(t)}{C(o)} = \left(1 + \frac{t}{\beta}\right)^{-(\alpha+1)} \tag{3.24}$$

and

$$\dot{U}(t) = \frac{\alpha+1}{\beta}\left(1 + \frac{t}{\beta}\right)^{-(\alpha+2)} = \frac{\alpha+1}{\beta}\, U^{\gamma} \tag{3.25}$$

where

$$\gamma = (\alpha+2)/(\alpha+1)$$

Thus the "lump" appears to decay as a γth order reactant with γ depending on the parameter α in the initial distribution. It is not surprising that as $\alpha \to \infty$, $\gamma \to 1$ for the variance σ^2 of the initial distribution is $(\alpha+1)/\beta^2$ and the mean, $1/(\alpha+1)\mu$ is $(\alpha+1)\beta$. Thus $\alpha^2/\mu^2 = 1/(\alpha+1) \to 0$ as $\alpha \to \infty$, the distribution becomes narrower and therefore appears to decay in first order fashion. It is remarkable that for all α and β the rate constant in eqn. (3.25) is $\mu = (\alpha+1)/\beta$, the mean value of k in the initial distribution. If $\alpha = 0$ the apparent order is $\gamma = 2$ and it is noteworthy that second order reaction rates have been used to correlate hydrocarbon cracking for some time.

Though it is seldom possible to get complete results of this kind, Luss exploited the convexity of the exponential to show that

$$e^{-\mu t} < U(t) < \left[\sigma^2 + \mu^2 e^{-\nu t}\right] \Big/ \left[\sigma^2 + \mu^2\right],$$

$$\nu = (\sigma^2 + \mu^2)/\mu.$$

Such a result is extremely useful (and incidentally an excellent illustration of the value of the theory of inequalities) as it gives bounds on an observable in terms of certain calculable functionals, in this case the mean rate constant and their variance in the initial distribution. For an extended discussion of continuous mixtures see Gavalas and Aris [73], while for some alternative treatments Liu and Lapidus [118] and Bailey [21] may be consulted.

4 How should a model be manipulated into its most responsive form?

"You boil it in sawdust: you salt it in glue:
You condense it with locusts and tape:
Still keeping one principal object in view--
To preserve its symmetrical shape."

C. L. Dodgson. The Hunting of the Snark.
Fit 5. St. 24.

4.1 Introductory suggestions.

Though a model may have been formulated with perfect propriety and perspicacity it is almost always a mistake to jump in with an extensive series of computations. It is better to live with it for a bit, to view it from different angles, to shape and mould it more justly. If the analogy may be permitted, there is a need for mathematical foreplay if model is to be fully responsive and the ultimate knowledge is to be satisfactory. The analogy is not inappropriate in that a like delicacy and tentativeness are required, but it breaks down in the bibliography for while the literature of the one art is enormous, though for the most part preternaturally dull, that of the other, for all its excitement, is thin on the ground. Nor is the maxim of going to the masters as much help as it might be, for though, like Jacob's, their wrestling is before the breaking of the day, unlike him, they seldom show it in their gait. There is of course a considerable literature on intellectual creativity but that is not the issue here. Our needs are closer to those that have been met by Polya's examination of the art of problem solving [145-147] but in some ways are even preparatory to this.

Lin and Segel's excellent text is one of the very few that addresses itself to this question, though there is much to be learned from Hammersley's vivacious writing on the strategy and tactics of research in mathematics [80-82] and Noble's book [136].

Hammersley's "maxims for manipulators" [80] are worth summarizing since they give the flavour of his thinking and raise several points we shall return to later. They are: i, clean up the notation; ii, choose suitable units; iii, reduce the number of variables wherever possible; iv, draw rough sketches and examine particular cases; v, avoid rigour like the plague, it only leads to rigor mortis at the manipulative stage; vi, have about an equal amount of stuff on each side of the equation. These he illustrates with a maximimization problem of deceptively innocent aspect. If this is tackled bullheadedly without the kind of prior manipulation Hammersley discusses it proves very resistant to accurate and apprehensible solution. Though he is obviously the poorer golfer brandishing a bigger bag of clubs, Aris' "maxims for mathematical modelling" [15] will perhaps bear repetition.

Maxims for Mathematical Modelling

1. Cast the problem in as elegant a form as possible.
2. Choose a sympathetic notation, but don't become too attached to it.
3. Make the variables dimensionless, since this is the only way in which their magnitudes take on general significance, but do not lose sight of the quantities which may have to be varied later on in the problem nor forget the physical origin of each part.

4. Use a priori bounds of physical or mathematical origin to keep all variables of the same order of magnitude, letting the dimensionless parameters show the relative size of the several terms.
5. Think geometrically. See when you can reduce the number of variables (even at the expense of first treating an over-simplified problem), but keep in mind the needs of the general case.
6. Use rough and ready methods, but don't carry them beyond their point of usefulness. (e.g. Isoclines in the phase plane.)
7. Find critical points and how the system behaves near them or what is asymptotic behaviour is at long or short times.
8. Check limiting cases and see how they tie in with simpler problems that can be solved explicitly.
9. Use crude approximations, e.g. 1-point collocation. Trade on the analogies they suggest, but remember their limitations.
10. Rearrange the problem. Don't get fixed ideas on what are the knowns and what the unknowns. Be prepared to work with implicit solutions.
11. Neglect small terms, but distinguish between regular and singular perturbations.
12. Use partial insights and despise them not. (e.g. Descartes rule of signs).
13. These maxims will self-destruct. Make your own!

Of these the most important is surely the last for they are not to be regarded as a book of rules with promise of success to any who will apply them. Rather they are suggestions and if anyone is to "think in the marrow bone" he must grow his own bone marrow.

The graith of the compleat mathematical modeller is a rich one and it is beyond the scope of these notes to survey the whole range of it, but we shall want to say something about the preliminaries of nomenclature and the choice of dimensionless variables. The possible reduction of the number of equations also deserves some attention, but the nub of the matter lies in the whole business of getting a feel for the solution without doing the detailed calculations. To see this done skillfully as, for example, by Paul Fife in his treatment of pattern formation [65], is to share the delight of the true craftsman. There are many tools that are available for this task and we shall only be able to touch on a few of them in any detail. However a preliminary list (in no particular order) would certainly include:

1. Obtaining a priori bounds on the solution;
2. Use of fixed point theorems and contraction maps;
3. Location of roots (Descartes' rule of signs);
4. Use of special properties (e.g. the constant cross-ratio of four solutions of Ricatti's equation);
5. Method of isoclines;
6. Degree theory;
7. Use of perturbation theory;
8. Asymptotic analysis, singular perturbation theory;
9. Group theory;
10. Use of inequalities;
11. Finding moments of the solution;
12. Intermediate asymptotics, travelling waves;
13. Integration by parts (e.g. Frisch's treatment of diffusion lag);
14. Maximum principles;

15. Numerical approximations (even crude ones like one-point collocation);
16. Variational principles;
17. Linearization;
18. Integral methods (e.g. in boundary layer theory).

4.2 Natural languages and notations.

Certain branches of mathematics lend themselves naturally to expressing the structure of a physical theory. It is well-known that the theory of relativity found a natural language in the calculus of tensors, or "Ricci-calcul" as it was first called. Similarly the language of Cartesian vectors and tensors is natural for mechanics and linear algebra for the stoichiometry of chemical reactions. Thus the physical realities are clearly represented by the vectorial equation $m\ddot{\underset{\sim}{r}} = f$ and though it may be necessary to use the equations in component form, $m\ddot{\underset{\sim}{x}} = f_x$, etc., one should never lose sight of the whole in dealing with the parts. The notational convention of having one symbol for the vector or tensor helps in this. When a component notation is used such rules as the summation convention are also valuable and the compactness of $\underset{\sim}{l}.\underset{\sim}{x}$ or $l_i x^i$ has advantage over $l_1x_1 + l_2x_2 + l_3x_3$ or $lx + my + nz$.

Hammersley's first maxim (to clean up the notation) has more to it than meets the eye, for the right choice of notation, though in some respects a matter of personal taste, is not to be dismissed as trivial. It is easy enough to make mistakes in manipulation under the best of circumstances but to burden oneself with a cumbersome and unsuitable notation is downright stupid. Except where some overriding convention has a prerogative, the system for the basic physical quantities should be generally as simple and mnemonic as possible. It is not always possible to avoid suffixes but again it is important that they should be natural. Thus in the example of App. C by calling the species $A_1, A_2 \ldots A_S$, it is natural to denote their concentrations by c_j, $j = 1,2 \ldots S$. Similarly at the early stage it is better to call the volume of the coolant V_c than to preempt yet another letter. How sympathetic one person's notation may appear to another is a matter of context and personal taste. For example chemical engineering texts often use Θ for the residence or holding time of the stirred tank, V/q, while biochemical engineers use D for its reciprocal, calling it the dilution rate; yet, since D plays the role of a death rate, many would feel that Θ itself would be preferable. In choosing symbols for the dimensionless variables it is seldom possible to make a consistent translation between Latin and Greek letters, though it is an admirable tradition to follow. Thus coordinates (x,y,z) can often become (ξ,η,ζ) or time t become dimensionless time τ. Both alphabets tend to run out all too quickly, but it is better to have recourse to Ps. 119 in the King James than to succumb to such symbolic solecisms as V for a diffusion coefficient or x for a mass. In general single letters should be used for a single quantity, but tradition has sanctified the union of a capital and lower case for the notable dimensionless groups such as the Reynolds number, Re. This seldom causes confusion and is less unsightly than the convention N_{Re}. Chemical engineers seem to have led the field in coining names for dimensionless groups and hence immortalizing one another--much as the naturalists of the 18th centruy did with the taxonomy of Linnaeus--though it may be questioned whether it helps to perpetuate Damköhler's memory in four numbers with Roman numeral suffixes. Referring again to the example of App. C, the concentrations and temperatures, c_j, T, T_w and T_c become u_j, v, w and Θ in their dimensionless forms. The last choice is not altogether a happy one, but v_c has to be avoided if we are to have ultimate freedom from suffixed for the dependent variables. In the case of Σ_6 whose reduction to dimensionless form is to be studied in detail in the next section we shall use v_c for the dimensionless coolant feed temperature. This however is a fixed constant, not a dependent variable, and we need Θ as the conventional symbol for residence time.

4.3 Rendering the variables and parameters dimensionless.

If the matter of notation is incidental, being governed by the canons of taste and common sense rather than high principle, the matter of making the variables dimensionless is of the essence. A physical magnitude has meaning with respect to an arbitrary set of standards and two quantities of the same dimensions measured in the same units can be compared. But even the world-wide adoption of SI units cannot give it any intrinsic meaning and the current effort in metricization, while useful to trade and engineering, has considerably less to contribute to science than would accrue from the revival of Latin as a lingua franca. It is only when quantities are made dimensionless that their magnitudes acquire an intrinsic meaning in the context of the model.

It might be worth remarking parenthetically that we are here practising so-called "inspectional analysis" [259,260], that is, the manipulation of the basic equations to reveal the dimensionless groups. It is the subtlest and most penetrating way of the methods sometimes called dimensional. Dimensional analysis, as expounded by Bridgeman [203] and others, relies on Rayleigh's method of indices [236] and the Buckingham Pi theorem to excogitate the dimensionless groups from a list of the significant physical quantities. The Pythagorean method, to use the name for the method of ratios coined by Becker in his excellent little monograph [201], refers to the technique of forming dimensionless ratios from the significant physical quantities peculiar to the system. Becker distinguishes three subgroups of dimensionless ratios: ratios of fluxes (e.g. of momentum energy or matter), which are appropriately called Heraclitian; geometrical ratios (e.g. aspect ratio), called the Pythagorean subgroup; and ratios of amounts or charges, called Democritian in recognition of the fact that the conceptual separation of matter from space goes back to Democritus and the atomists. He uses the term "configuration analysis" for the incorporation of the Pythagorean method into a larger analysis of physical systems and claims that it is "immensely more persuasive than dimensional analysis, almost as penetrating as inspectional analysis, and more versatile than either of these." The whole subject of dimensional analysis has a considerable literature; cf. [202,203,207-9,212,215-8,220-1,229-33,236-9,247,259,260].

The principle of making the equations dimensionless is simple enough. Each variable, dependent or independent, is expressed as a product of some characteristic quantity of the same dimensions and a dimensionless variable. The equations are then rearranged until a suitable set of dimensionless parameters appears. Two principles govern this process: i, constant quantities should be used as characteristic quantities; ii, the dimensionless parameters should bear the burden of showing the comparative importance of the various terms in the equation.

Let us try to elucidate this by considering the model Σ_6 of App. C the stirred tank with a single first-order irreversible reaction. There are usually several different ways of rendering equations non-dimensional and of choosing the dimensionless parameters and their pros and cons are best illustrated by an example. If we put $k(T) = A\exp(-E/RT)$ in eqns. (C15) and (C16) and replace hA in (C16) by h (to avoid confusion between the area and the pre-exponential factor in k), the equations of this model are:

$$V\frac{dc}{dt} = q(c_f - c) - VAe^{-E/RT}c \tag{4.1}$$

$$VC_p\frac{dT}{dt} = qC_p(T_f - T) - h(T - T_{cf}) + (-\Delta H)VAe^{-E/RT}c \tag{4.2}$$

Note that the same letters will be used for the same kind of parameters in the different forms of the equation but they will be differently related to the physical constants in the different modes of non-dimensionalizing. They are always defined in context.

Let c^*, T^* and t^* be a characteristic concentration, temperature and time to be chosen later and

$$\tau = t/t^*,\ u = c/c^*,\ v = T/T^* \tag{4.3}$$

and, by the persistence and simplification of indices,

$$u_f = c_f/c^*,\ v_f = T_f/T^*,\ v_c = T_{cf}/T^*. \tag{4.4}$$

Multiplying the first equation by t^*/Vc^* gives

$$\frac{du}{d\tau} = \frac{qt^*}{V}(u_f - u) - At^*ue^{-E/RT^*v} \tag{4.5}$$

There are three parameters qt^*/V, At^* and E/RT^*. Now V/q is the so-called residence time so that V/qt^* is the dimensionless residence time, say Θ, At^* is a measure of the reaction rate of Damköhler number, say α, and E/RT^* is the dimensionless activation energy or Arrhenius number, say γ. Thus

$$\frac{du}{d\tau} = \frac{u_f - u}{\Theta} - \alpha ue^{-\gamma/v} \tag{4.6}$$

The second equation must be multiplied by t^*/VC_pT^* to give

$$\frac{dv}{d\tau} = \left[\frac{qt^*}{V}v_f + \frac{ht^*}{VC_p}v_c\right] - \left[\frac{qt^*}{V} + \frac{ht^*}{VC_p}\right]v + \frac{(-\Delta H)c^*}{C_pT^*}At^*ue^{-\gamma/v}$$

From this it is clear that $(-\Delta H)c^*/C_pT^* = \beta$, say, is an important parameter; moreover it has the nature of a dimensionless temperature rise due to reaction for $(-\Delta H)c^*$ is the heat released by the complete reaction of the feed at the characteristic concentration. For the moment let us not reduce everything completely but set

$$\beta = \frac{(\Delta H)c^*}{C_pT^*}, \; \delta_1 = \frac{qt^*}{V} = \frac{1}{\theta}; \; \delta_2 = \frac{ht^*}{VC_p}, \; \delta = \delta_1 + \delta_2 \tag{4.7}$$

Then

$$\frac{dv}{d\tau} = (\delta_1 v_f + \delta_2 v_c) - \delta v + \alpha\beta u e^{-\gamma/v} \tag{4.8}$$

$$= \delta(\bar{v}-v) + \alpha\beta u e^{-\gamma/v}$$

if we define

$$\bar{v} = (\delta_1 v_f + \delta_2 v_c)/\delta \tag{4.9}$$

as the weighted mean of the feed and coolant temperatures.

We now have various directions in which we can go. First suppose that all conditions are to be held constant and that we are interested in reducing the number of parameters to the minimum. Then in the equation for u we can make u_f, γ and θ all equal to 1 by choosing

$$c^* = c_f, T^* = E/R, \; t^* = V/q \tag{4.10}$$

giving

$$\frac{du}{d\tau} = 1 - u - \alpha u e^{-1/v} \tag{4.11}$$

No amount of manipulation will get rid of α. In the equation for v there appear to be five additional constants δ_1, δ_2, v_f, v_c and β, but $\delta_1 = 1/\Theta = 1$ since all conditions are constant v_c and v_f can be condensed into $\bar{v}$ by eqn. (4.9). It is well to take one constant over to the left since this will be eliminated when we set the derivatives equal to zero to study the steady state. Thus let $\bar{v}$ be defined as above and

$$\zeta = \beta/\delta \tag{4.12}$$

so that

$$\frac{1}{\beta}\frac{dv}{d\tau} = \frac{\bar{v}-v}{\zeta} + \alpha u e^{-1/v} \tag{4.13}$$

Between equations (4.11) and (4.13) there are four parameters

$$\alpha = Av/q,\ \beta = (-\Delta H)Rc_f/C_pE,\ \zeta = [((-\Delta H)Rc_f/EC_p]/[1+h/qC_p] \tag{4.14}$$

and

$$\bar{v} = \frac{R(qC_pT_f+hT_{cf})}{E(qC_p+h)} \tag{4.15}$$

The steady state equations however contain only three, α, ζ and $\bar{v}$. They can be combined into a single equation by substituting for u from eqn. (4.11) in eqn. (4.13),

$$v-\bar{v} = \zeta - \frac{\alpha e^{-1/v}}{1+\alpha e^{-1/v}} \tag{4.16}$$

This equation is studied later in Sec. 4.5. The points to be made in favour of this mode of non-dimensionalizing are that it does give us the simplest possible form of the equations, for example, there is no parameter in the exponential function. The objection that can be raised to it is that the magnitudes of u and v are generally very different; u is between 0 and 1 but if E/R is of the order of 10000°, as it may well be, even a fairly high temperature such as 500°K makes v of the order of 0.05.

A second method--and one that overcomes this difficulty--is to let u_f, v_f and Θ all be equal to 1, by putting

$$c^* = c_f,\ T^* = T_f,\ t^* = V/q. \tag{4.17}$$

Then

$$\frac{du}{d\tau} = 1 - u - \alpha u e^{-\gamma/v} \tag{4.18}$$

$$\frac{dv}{d\tau} = (\delta_1 + \delta_2 v_c) - \delta v + \alpha\beta u e^{-\gamma/v} \tag{4.19}$$

We thus have

$$\alpha = AV/q, \ \beta = (-\Delta H)c_f/C_p T_f, \ \gamma = E/RT_f, \ \delta_1 = 1 \tag{4.20}$$

and

$$\delta_2 = h/qC_p$$

If T_f and T_c are not too disparate it is advantageous to put

$$T^* = (qC_p T_f + hT_{cf})/(qC_p + h) \tag{4.21}$$

since then the second equation becomes

$$\frac{1}{\delta}\frac{dv}{d\tau} = 1 - v + \zeta\alpha u e^{-\gamma/v} \tag{4.22}$$

where

$$(1/\delta) = qC_p/(qC_p + h), \ \zeta = qc_f(-\Delta H)/(qC_p T_f + hT_{cf}) \tag{4.22}$$

A variant of the second method considers u and v to be the deviations of c and T from c_f and T_f by setting

$$u = (c_f - c)/c_f, \ v = (T - T_f)/T^*. \tag{4.23}$$

Before choosing T^* let us see what this does to the exponential function. Since $T = T_f + T^*v$, $\exp-(E/RT) = \{\exp-(E/RT_f)\}\{\exp E(T-T_f)/RTT_f\} = \exp-(E/RT_f)\exp\{ET^*v/RT_f^2(1+T^*v/T_f)\}$. Thus if we choose $T^* = RT_f^2/E$ and put $\gamma = E/RT_f$ we have

$$k(T) = Ae^{-E/RT} = k(T_f)\exp[v/(1+v/\gamma)]. \tag{4.24}$$

The choice has two points in its favour. In the first place, it brings $k(T_f)$ out as a factor and this with c_f gives $k(T_f)c_f$ which is the reaction rate at feed conditions. In fact the equations are now:

$$\frac{du}{d\tau} = -u + \alpha(1-u)\exp[v/(1+v/\gamma)], \quad (4.25)$$

$$\frac{1}{\delta}\frac{dv}{d\tau} = -v + v_c + \alpha\zeta(1-u)\exp[v/(1+v/\gamma)], \quad (4.26)$$

where $\alpha = Vk(T_f)/q$, $\beta = (-\Delta H)c_f E/C_p RT_f^2$, $\gamma = E/RT_f$, $\delta = (qC_p + h)/qC_p$, $\zeta = \beta/\delta$ and $V_c = hE(T_{cf}-T_f)/RT_f^2(h + qC_p)$. Again, if the weighted mean feed temperature $(qC_pT_f + hT_{cf})/(qC_p + h)$ were used everywhere in place of T_f, we would have the same equations but $V_c = 0$. The second advantage of this form is that it makes natural the approximation of the exponential function by e^v, an approximation which can be obtained formally by letting $\gamma \to \infty$.

Finally--for it would be possible to play a large number of variations on this simple theme--consider the situation when an important physical quantity that affects more than one parameter, for example, q, is to be varied. If neither limiting case, $q \to 0$ or $q \to \infty$, is to be considered there is no real objection to retaining $t^* = V/q$ since the equations are autonomous. If however we wish to avoid this we can take $t^* = VC_p/h$ giving $\theta = h/qC_p$, $\alpha = AVC_p/h$, $\delta_1 = 1/\theta$, $\delta_2 = 1$. The reference temperature should not be taken to be the weighted mean as in eqn. (4.21) since this will vary with q but we can take $v = E(T-T_f)/RT_f^2$. Thus, for example, with

$$u = (c_f-c)/c_f, \; v = E(T-T_f)/RT_f^2, \; v_c = E(T_{cf}-T_f)/RT_f^2, \quad (4.27)$$

$$\alpha = AVC_p/h, \; \beta = (-\Delta H)Ec_f/C_pRT_f^2 \quad (4.28)$$

we have

$$\theta\frac{du}{d\tau} = -u + \alpha\theta(1-u)\exp[v/(1+v/\gamma)] \quad (4.29)$$

$$\theta\frac{dv}{d\tau} = \theta v_c - (1+\theta)v + \alpha\beta\theta(1-u)\exp[v/(1+v/\gamma)]. \quad (4.30)$$

In this form the equations have parameters α, β, γ, v_c and Θ, the last being the variable parameter. This is the case extensively studied by Uppal, Ray and Poore [184], though they took a slightly different dimensionless time the effect of which was to confine the region of multiplicity to $0 < \Theta < 1$.

Another question that may be raised about any particular choice of dimensionless variables is whether it fits in with a larger scheme. For example if we replace T_{cf} in eqn. (4.2) by T_c and append a third equation of the form of (C 14)

$$v_c C_{pc} \frac{dT_c}{dt} = q_c C_{pc}(T_{cf}-T_c) + h(T-T_c) \qquad (4.31)$$

we have three equations for c, T and T_c. Consider the extension of the form of eqns. (4.25) and (4.26) by setting

$$u = (c_f-c)/c_f,\ v = (T-T_f)E/RT_f^2,\ w = (T_c-T_f)E/RT_f^2 \qquad (4.32)$$

with

$$t^* = V/q,\ \alpha = Vk(T_f)/q,\ \beta = E(-\Delta H)c_f/RT_f^2C_p,\ \gamma = E/RT_f,$$
$$\delta_1 = 1,\ \delta_2 = h/qC_p, \omega = V_cC_{pc}/VC_p,\ \chi = q_cC_{pc}/qC_p. \qquad (4.33)$$

Then we have

$$\frac{du}{d\tau} = -u + \alpha(1-u)\exp[v/(1+v/\gamma)], \qquad (4.34)$$

$$\frac{1}{\delta}\frac{dv}{d\tau} = -v + \frac{\delta_2}{\delta} w + \alpha\beta(1-u)\exp[(1+v/\gamma)], \qquad (4.35)$$

$$\omega \frac{dw}{d\tau} = \chi(w_f-w) + \delta_2(v-w), \qquad (4.36)$$

a set of three equations with two additional parameters ω and χ the static and fluent heat capacity ratios.

A similar example which we will take up from other aspects later can be given rather briefly at this point. It concerns the growth of two organisms in a chemostat--the microbiologist's name for a stirred tank reactor. Two organisms, species 1 and 2, are present in concentrations, X_1 and X_2 and feed on a common nutrient whose concentration is S in a vessel whose volume is $(1/D)$ times the flow rate, i.e. $q = DV$. (This is the standard notation of biochemical engineering and D is known as the dilution rate). They grow at rates μ_1 and μ_2 which are functions of S and one moiety of S yields Y_1 and Y_2 moieties of the two species respectively. Thus

$$\frac{dX_1}{dt} = \mu_1(S)X_1 - DX_1 \tag{4.37}$$

$$\frac{dX_2}{dt} = \mu_2(S)X_2 - DX_2 \tag{4.38}$$

$$\frac{dS}{dt} = \frac{\mu_1(S)X_1}{Y_1} - \frac{\mu_2(S)X_2}{Y_2} + D(S_F - S) \tag{4.39}$$

where only nutrient is fed to the chemostat and at a concentration S_F. The functions $\mu_i(S)$ are of the form

$$\mu_i(S) = M_i \left\{1 + \frac{K_i}{S} + \frac{S}{L_i}\right\}^{-1} \tag{4.40}$$

We want to investigate the way in which the system depends on S_F and D so it will not do to use these values to make the concentrations and time dimensionless. Without loss of generality we can give the label 1 to the species which has the greater growth rate for large S, i.e. $M_1L_1 \geq M_2L_2$. Then $1/M_1$ has the dimensions of time and K_1 of concentration and accordingly we let

$$x = X_1/Y_1K_1,\quad y = X_2/Y_2K_1,\quad z = S/K_1,\quad \tau = M_1t. \tag{4.41}$$

There are then four fixed parameters

$$\alpha = M_2/M_1, \quad \beta = K_2/K_1, \quad \gamma = L_2/K_1, \quad \delta = L_1/K_1 \tag{4.42}$$

and two we wish to vary

$$\Theta = D/M_1, \quad Z = S_F/K_1. \tag{4.43}$$

Then

$$\begin{aligned} x &= \{g_1(z) - \theta\}x \\ y &= \{g_2(z) - \theta\}y \\ z &= \theta(Z-z) - xg_1(z) - yg_2(z) \end{aligned} \tag{4.44}$$

where

$$g_1 = \left\{1 + \frac{1}{z} + \frac{z}{\delta}\right\}^{-1}, \quad g_2 = \alpha\left\{1 + \frac{\beta}{z} + \frac{z}{\gamma}\right\}^{-1} \tag{4.45}$$

We remark that γ and δ can be infinite if there is no substrate inhibition.

4.4 Reducing the number of equations and simplifying them.

The example we have been considering extensively in the last section was a model, Σ_6, with just two equations. It had gotten to be that way by first reducing the model to one described by $S+1$ equations, (one for the concentration of each species and one for the temperature) and then observing that, since the reaction rate depended on only one of the concentrations, only one equation was needed in the concentration and one in the temperature. However for constant feed conditions a similar reduction can be made for the completely general reaction $\Sigma\alpha_j A_j = 0$, with reaction rate $r(c_1, c_2 \ldots c_S, T)$. For substitute

$$c_j = (q_f c_{jf}/q) + \alpha_j \xi \tag{4.46}$$

in the equation

$$V \frac{dc_j}{dt} = q_j c_{jf} - qc_j + \alpha_j Vr(c_1, \ldots c_S, T) \tag{4.47}$$

and α_j comes out as a factor leaving

$$V \frac{d\xi}{dt} = -q\xi + Vr(\xi, T) \tag{4.48}$$

where

$$\begin{aligned} r(\xi,T) &= r(\xi,T;\ q_1,\ldots q_r,\ c_{1f}\ldots c_{Sf}) \\ &= r((q_1 c_{1f}/q) + \alpha_1\xi,\ldots,T) \end{aligned} \tag{4.49}$$

Thus a single equation has been obtained for a single reaction. But it will be objected that the initial values $c_j(0) = c_{jo}$ may not all be expressible in the form

$$c_{jo} = (q_j c_{jf}/q) + \alpha_j \xi_o. \tag{4.50}$$

If they can all be so expressed then there is no problem for ξ_o then becomes the initial value for eqn. (4.48), but if they cannot then at least two values of

$$\eta_j = \{c_{jo} - (q_j c_{jf}/q)\}/\alpha_j \tag{4.51}$$

are different. Then let

$$c_j = (q_j c_{jf}/q) + \alpha_j(\xi + \eta_j\zeta) \tag{4.52}$$

a form which can take care of the initial conditions by letting $\xi(0) = 0$, $\zeta(0) = 1$. Substituting eqn. (4.52 in eqn. (4.38) and dividing through by α_j gives

$$\left[V \frac{d\xi}{dt} + q\xi - Vr(\xi,\zeta,T)\right] + \eta_j \left[V \frac{d\zeta}{dt} + q\zeta = 0\right] \tag{4.53}$$

Now each of the expressions in brackets must be separately zero for otherwise all the η_j would be the same. The first bracket gives eqn. (4.48) except that now the reaction rate is a function of ζ as well as of ξ and T and, parametrically, of c_{jo} as well as of c_{jf} and q_j. However the equation

$$V \frac{d\zeta}{dt} = -q\zeta \tag{4.54}$$

has the immediate solution $\zeta(t) = \exp - qt/V$ and quickly tends to zero. We say that the feed and initial compositions are compatible if one can be derived from the other by some degree of reaction, i.e. they are related by eqn. (4.50). The ζ is a measure of the incompatibility of the current composition and the feed composition and it is very satisfying to see that this just "washes out" of the reactor by a purely physical equation without any reactive term as the memory of an initial composition should.

In a more general situation there might be R simultaneous reactions

$$\sum_{j=1}^{S} \alpha_{ij}A_j = 0, \quad i = 1,2,\ldots R, \tag{4.55}$$

which without loss of generality we can take to be independent, i.e. the RxS matrix with entries α_{ij} is of rank R. Then the mass balance over the jth species gives

$$V \frac{dc_j}{dt} = q(\bar{c}_{jf} - c_j) + V \sum_{i=1}^{R} \alpha_{ij}r_i(c_1,\ldots c_S,T) \tag{4.56}$$

where $q\bar{c}_{jf} = q_j c_{jf}$. The substitution

$$c_j = \bar{c}_{jf} + \sum_{i=1}^{R} \alpha_{ij}\xi_i + (c_{jo} - \bar{c}_{jf})\zeta \tag{4.57}$$

gives

$$\sum_{i=1}^{R} \alpha_{ij} \left[V \frac{d\xi_i}{dt} + q\xi_i - Vr_i(\xi_1,\ldots \xi_R,\zeta T)\right]$$

$$+ (c_{jo}-\bar{c}_{jf})\left[V \frac{d\zeta}{dt} + q\zeta\right] = 0.$$

The independence of the reactions and the incompatibility of the feed and initial compositions force each of the bracketed expressions to be severally zero and so give $R+1$ equations (or R, if the feed and initial compositions are compatible) in place of S.

When the reactor is adiabatic, a combination of concentration and temperature obeys a similarly simple equation. For, taking $h=0$ in eqn. (4.20) gives $\delta = \delta_1 = 1$, $\delta_2 = 0$, so that multiplying eqn. (4.18) by and adding eqn. (4.19) gives

$$\frac{d}{d\tau}(\beta u+v) = (1+\beta) - (\beta u+v). \tag{4.58}$$

This notion can be carried over to distributed systems as when the reaction $\Sigma\alpha_j A_j = 0$ takes place at a rate r per unit volume in a porous pellet, Ω, through which the reactants diffuse with effective Knudsen diffusion coefficients D_j. For this system we have the equations

$$D_j \nabla^2 c_j + \alpha_j r(c_1,\ldots,T) = 0 \text{ in } \Omega \tag{4.59}$$

with

$$c_j = c_{js} \text{ on } \partial\Omega. \tag{4.60}$$

This implies that each linear combination $(D_j c_j/\alpha_j) - (D_k c_k/\alpha_K)$ satisfies

$$\nabla^2\{(D_j c_j/\alpha_j) - (D_k c_k/\alpha_k)\} = 0 \text{ in } \Omega \tag{4.61}$$

and is constant over $\partial\Omega$. But a potential function constant on $\partial\Omega$ is constant everywhere in Ω and hence we can substitute

$$c_j = c_{js} + (\alpha_j/D_j)\xi \tag{4.62}$$

to give

$$\nabla^2\xi + r(c_{1s} + (\alpha_1/D_1)\xi,\ldots,T) = 0 \text{ in } \Omega, \tag{4.63}$$

with

$$\xi = 0 \text{ on } \partial\Omega.$$

The energy balance gives

$$k_e \nabla^2 T = (\Delta H) r(c_1, \ldots T) \text{ in } \Omega$$

with (4.64)

$$T = T_s \text{ on } \partial\Omega$$

and this suggests the substitution

$$T = T_s + (-\Delta H/k_e)\xi. \tag{4.65}$$

Thus all the equation collapse into one, namely

$$\nabla^2 \xi = -R(\xi) = -r(c_{1s} + (\alpha_1/D_1)\xi, \ldots, T_s + (-\Delta H/k_e)\xi) \text{ in } \Omega$$

with (4.66)

$$\xi = 0 \text{ on } \partial\Omega.$$

Notice, however, that this is of restricted application. Except for symmetric regions, such a sphere, it does not apply with the more general boundary conditions

$$D_j \frac{\partial c_j}{\partial n} = k_j(c_{jf} - c_j) \text{ on } \partial\Omega.$$

Nor can it be extended to transients except in the special case of $(D_j \rho c_p / k_e) = 1$ for all j. In this case we write the transient equations as

$$\frac{\partial c_j}{\partial t} = D_j \nabla^2 c_j + \alpha_j r \;; \quad \rho c_p \frac{\partial T}{\partial t} = k_e \nabla^2 T + (-\Delta H) r \tag{4.67}$$

and assume the initial values are uniform and c_{jo}, T_o respectively. Then putting

$$c_j = c_{js} + (\alpha_j/D_j)\xi + (c_{jo} - c_{js})\zeta$$

(4.68)

$$T = T_s + (-\Delta H/k_e)\xi + (T_o - T_s)\zeta$$

and letting

$$D = D_j = k_e/\rho c_p$$

gives

$$\left[\frac{\partial \xi}{\partial t} - D\nabla^2\xi - R\right] + \frac{(c_{jo}-c_{js})D_j}{\alpha_j}\left[\frac{\partial \zeta}{\partial t} - D\nabla^2\zeta\right] = 0. \tag{4.69}$$

Also $\xi = \zeta = 0$ on $\partial\Omega$ and $\xi = 0$, $\zeta = 1$ initially. If all the quantities $(c_{jo}-c_{js})D_j/\alpha_j$ are the same then we can fix ξ_o and ignore ζ. But if they are different then the same argument as before shows that the two brackets are severally equal to zero. Thus ζ is the distribution of temperature in Ω which is initially uniform and has zero boundary values and such a temperature, as we know, subsides to zero everywhere.

When such a reduction can be made the stability picture is gained from the reactive equation and the intrinsically stable equation for ζ adds nothing to the analysis of stability. For example, if

$$\frac{du}{d\tau} = 1 - u - \alpha R(u,v)$$
$$\frac{dv}{d\tau} = 1 - v + \alpha\beta R(u,v) \tag{4.70}$$

and $w = \beta u + v$, then an equivalent system is

$$\frac{du}{d\tau} = 1 - u - \alpha R(u,w-\beta u)$$
$$\frac{dw}{d\tau} = 1 + \beta - w \tag{4.71}$$

The linearization of the first about a steady state (u_s,v_s) gives

$$\begin{bmatrix}\dot{x}\\ \dot{y}\end{bmatrix} = \begin{bmatrix}-(1+\alpha R_u) & -\alpha R_v\\ \alpha\beta R_u & -(1-\alpha\beta R_v)\end{bmatrix}\begin{bmatrix}x\\ y\end{bmatrix} \tag{4.72}$$

where $x = u-u_s$, $y = v-v_s$. Similarly, if $z = w-w_s$, the linearization of eqns. (4.71) gives

$$\begin{bmatrix} \dot{x} \\ \dot{z} \end{bmatrix} = \begin{bmatrix} -(1 + \alpha R_u - \alpha\beta R_v) & -\alpha R_v \\ \cdot & -1 \end{bmatrix} \begin{bmatrix} x \\ z \end{bmatrix} \tag{4.73}$$

The stability criteria for these two matrices are the same.

The example of competing organisms that we intend to take up again later can be done in the short order now. The addition of the three equations in (4.44) gives

$$\dot{x} + \dot{y} + \dot{z} = \Theta\{Z - (x+y+z)\} \tag{4.74}$$

Thus if the point (x,y,z) does not lie in the plane

$$x + y + z = Z \tag{4.75}$$

it rapidly approaches it since

$$x + y + z = Z + \{x_o + y_o + z_o - Z\}e^{-\Theta\tau}$$

From this we see that all the steady states must lie in the plane, ABC of the figure below. If M is a starting point in ABC the trajectory of

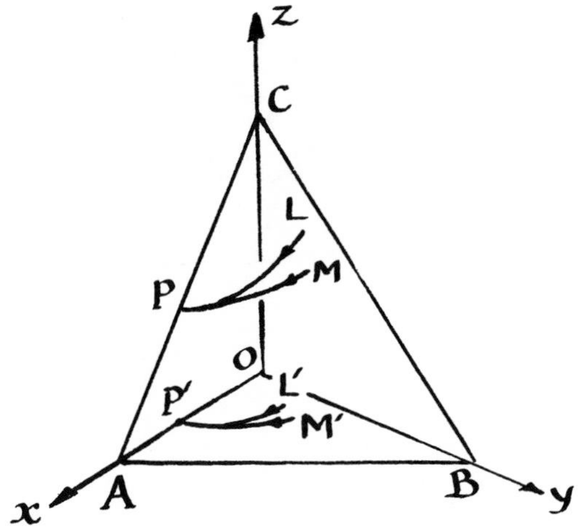

the solution of eqn. (4.44) lies wholly in the plane and approaches the steady state. The trajectory from a nearby point L, not in the plane, quickly approaches the trajectory MP and becomes tangential to it, if, as is the case shown here, the other eigenvalues are negative and greater than

Θ. This relationship is preserved in the projections L'P', M'P' on the plane OAB. It follows that we shall get the essence of the behaviour of the system by restricting attention to the plane

$$z = Z - x - y.$$

In this case we have the equations

$$\dot{x} = \{g_1(Z - x - y) - \Theta\}x$$

$$\dot{y} = \{g_2(Z - x - y) - \Theta\}y$$

For a certain combination of $\alpha, \beta, \gamma, \delta$, Θ and Z the phase plane of solutions of these two equations looks like this:

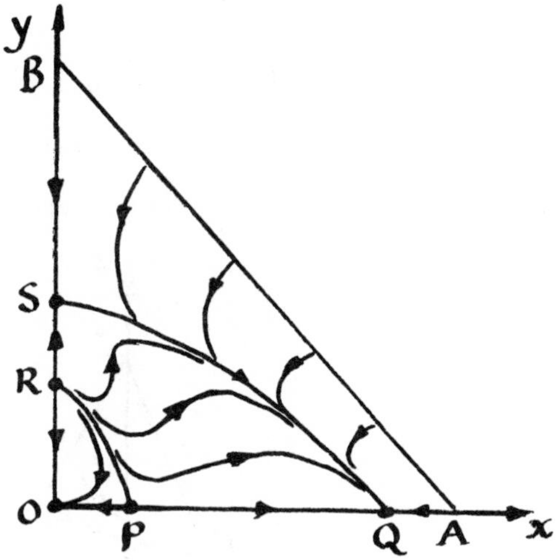

There are two stable steady states 0 and Q. Any initial state in the triangular region OPR is attracted to 0, while all trajectories starting outside, i.e. in PRBA, go to Q. Thus RP is a separatrix between the two regions of attraction. It is the projection of the curve RP in the plane ABC, shown in the next figure (N.B. it is not P"R") onto OAB. The surface P'R'RR"P"P is the separatrix surface in space and allows for an arbitrary starting point. Thus any trajectory starting between this surface and the z-axis goes to 0; all other trajectories go to Q. For some purposes we may need the whole separatrix surface but its structure is often sufficiently clear from its intersection with the plane.

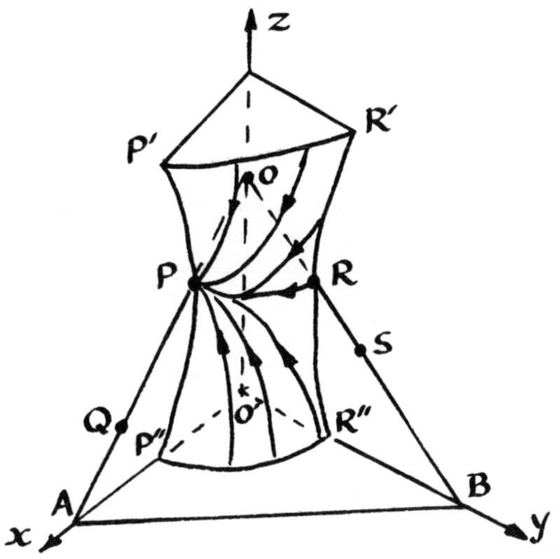

4.5 Getting partial insights into the form of the solution.

In the introductory paragraph to this chapter a number of techniques for getting some feel for the form of the solution were listed. This is not the place to explore them all in detail, but it is important to see some of them in action since they are at the heart of the whole craft of modelling. We shall do this by considering three examples. The first is an illustration of phase plane techniques in the chemostat that we have just been looking at; the second is the use of a coarse numerical method to suggest an approach to the uniqueness problem; the third is intended to show how far one can get toward a complete picture without any calculation.

4.5.1 The phase plane and competing populations.

The pair of equations

$$\dot{x} = \{g_1(Z - x - y) - \Theta\}x \qquad (4.76)$$

$$\dot{y} = \{g_2(Z - x - y) - \Theta\}y \qquad (4.77)$$

can have four types of steady state which can be obtained by setting $\dot{x} = \dot{y} = 0$. They bear the following interpretations and we shall use the notation shown at the left.

0	$x_s = y_s = 0$	complete wash-out of both x_1 and x_2
P or P and Q	$g_1(Z-x_s) = \Theta,\ y_s = 0$	x_1 grows, x_2 washes out
R or R and S	$x_s = 0,\ g_2(Z-y_s) = \Theta$	x_1 washes out, x_2 grows
T or T and U	$x_s+y_s=\zeta, g_1(Z-\zeta)=g_2(Z-\zeta) = \Theta$	both x_1 and x_2 coexist but in indeterminate proportions

The last possibility only arises if the curves $g_1(z)$ and $g_2(z)$ cross one another and then Θ must be chosen to have their common value. If the curves g_1 and g_2 are disposed as below and the point (Z,Θ) is at 0, then there are five steady states named according to the rule that proceeding to the left from 0 the first intersection with g_1 is P and the second, if there is one, is called Q; likewise the first and second intersections with g_2 are respectively R and S. Reading from right to left this disposition is uniquely described by the 'word' OPRSQ. In the x,y-plane this is also the order of increasing distance from the origin though P and Q are of course on the x-axis whilst R and S are on the y-axis. We can put the steady states in part (b) of the preceding figure. The 45° diagonals have been drawn in not only to confirm the order of placing the points but also with the reminder that they are isoclines of horizontal and vertical passage. Thus $\dot{x}=0$ on the diagonals through P and Q and $\dot{y}=0$ on

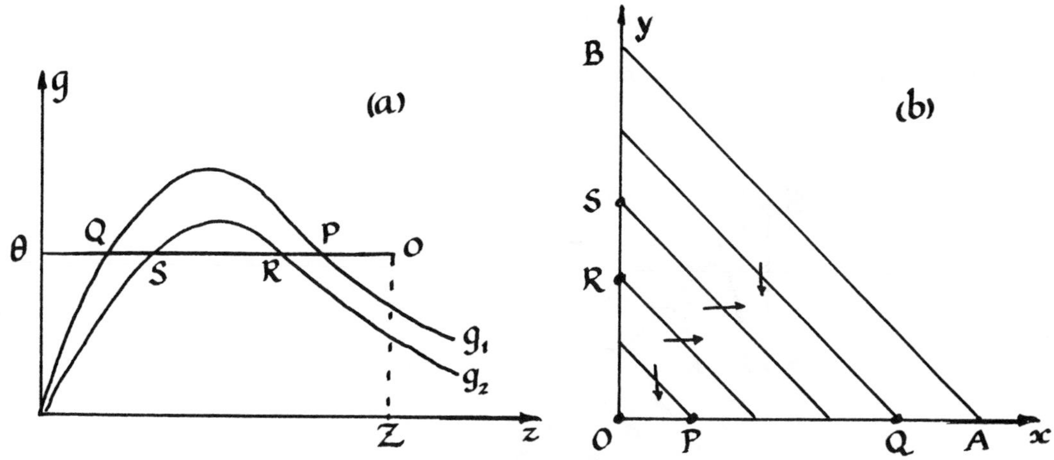

those through R and S. Moreover we can put arrows on the line segments for we see by part (a) of the figure that $g_2 < \Theta$ at P and so $\dot{y} < 0$ on the diagonal through P in part (b). The axes x=0 and y=0 are solutions of the equations and we can put arrows on the segments by glancing at part (a) of the figure. The other diagonals are not isoclines but we notice that on $x+y = Z-\zeta$

$$\frac{dy}{dx} = K\,\frac{y}{x} \; : \quad K = \frac{g_2(\zeta)-\Theta}{g_1(\zeta)-\Theta} \,,$$

that is the tangent of the angle that is the direction of the solution is a constant multiple of (y/x), the tangent of the direction from (x,y) to the origin. K = 1 on AB and zero or infinite on the diagonals through the steady states, and we can see that it varies roughly in this fashion:

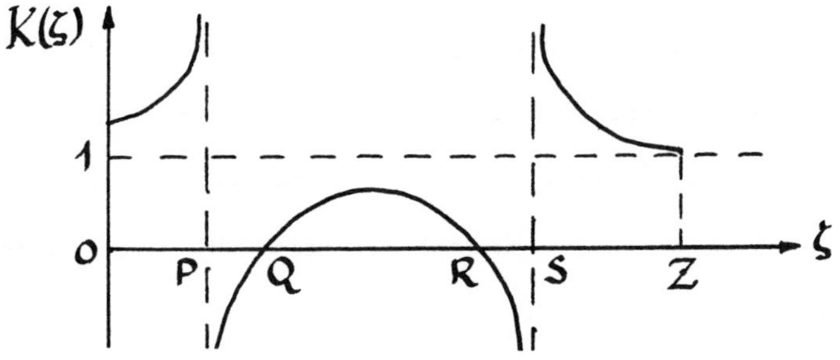

Thus in the trapezoidal region between the diagonals through R and S the flow must be northeasterly and the slope of the paths increase to a maximum and then decrease again to zero.

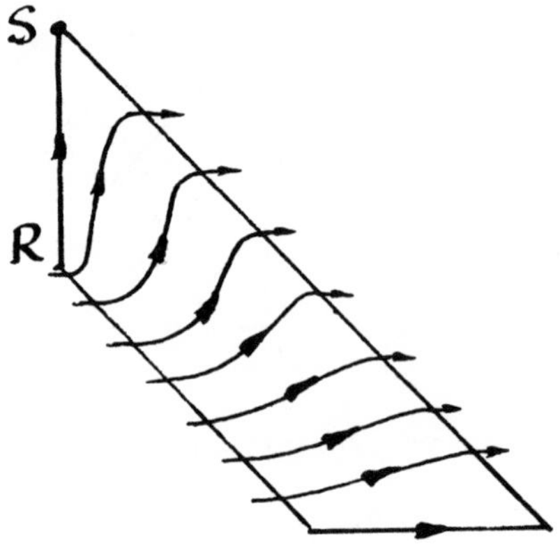

The direction of the solutions on AB is straight to the origin since $K = 1$ on AB. If this is done for each section a portrait emerges in which there will be south-south-easterly and east-south-easterly trajectories connecting R with P and S with Q in their respective trapezoids.

Thus we have already a general impression of the solution and see that RP will be a separatrix between the regions of attraction of 0 and Q the two stable steady states. We also see where computation may be difficult. For example, if it is important to be accurate in the neighbourhood of R the rapid change of direction may give trouble.

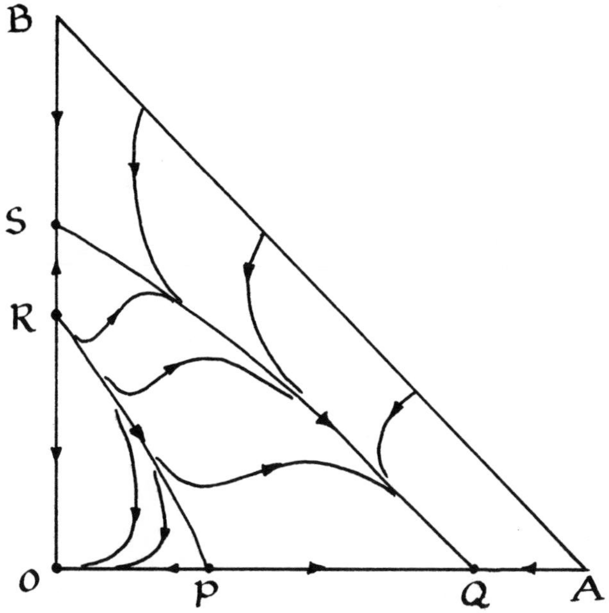

All this has been done without any computation. It can be confirmed (or perhaps eased over a sticking point) by linearization. Eqns. (4.76) and (4.77) can be linearized about any steady state in the form

$$\dot{\xi} = F\xi - G\eta$$
$$\dot{\eta} = -H\xi + K\eta \tag{4.78}$$

where

$$x = x_s + \xi,\ y = y_s + \eta \tag{4.79}$$

and

$$F = g_1(Z-x_s-y_s) - \Theta - x_s g_1'(Z-x_s-y_s)$$
$$G = x_s g_1'(Z-x_s-y_s)$$
$$H = y_s g_2'(Z-x_s-y_s) \tag{4.80}$$
$$K = g_2(Z-x_s-y_s) - \Theta - y_s g_2'(Z-x_s-y_s)$$

are constants since they are evaluated at the steady state under scrutiny. Because of the equations satisfied at the various steady states, the constants F,...K simplify as follows

Type of Steady State	F	G	H	K
0	$g_1 - \Theta$	0	0	$g_2 - \Theta$
P,Q	-G	G	0	$g_2 - \Theta$
R,S	$g_1 - \Theta$	0	H	-H
T,U	-G	G	H	-H

The next table shows what the eigenvalues and eigenvectors of the linearization are and what the conditions are that give the steady state its character. These characters are denoted by SN for stable node, UN for unstable node and SP for saddle point. The suffix on SP tells which eigenvector is tangent to the unique incoming trajectory. In the case of T and U one eigenvalue is always zero and the stability or instability of a point is governed by the sign of the other; these are denoted by SS and SU for semi-stable and semi-unstable.

From this table we can read the character of the steady states in the example OPRSQ above. D is a stable node, P a saddle point (SP_2), Q another stable node, R an unstable node and S an SP_2.

Steady State	Eigenvalues and vectors $\lambda_1, \underset{\sim}{v}_1$	$\lambda_2, \underset{\sim}{v}$	Character and conditions
0	$F, \begin{bmatrix}1\\0\end{bmatrix}$	$K, \begin{bmatrix}0\\1\end{bmatrix}$	SN, $g_1, g_2 < \Theta$ $SP_1, g_1 < \Theta < g_2$ $SP_2, g_2 < \Theta < g_1$ UN, $\Theta < g_1, g_2$
P,Q	$-G, \begin{bmatrix}1\\0\end{bmatrix}$	$K, \begin{bmatrix}-G\\K+G\end{bmatrix}$	SN, $G > 0,\ g_2 < \Theta$ $SP_1, G > 0,\ g_2 > \Theta$ $SP_2, G < 0,\ g_2 < \Theta$ UN, $G < 0,\ g_2 > \Theta$
R,S	$F, \begin{bmatrix}F+H\\-H\end{bmatrix}$	$-H, \begin{bmatrix}0\\1\end{bmatrix}$	SN, $H > 0,\ g_1 < \Theta$ $SP_1, H < 0,\ g_1 < \Theta$ $SP_2, H > 0,\ g_1 > \Theta$ UN, $H < 0,\ g_1 > \Theta$
T,U	$0, \begin{bmatrix}-1\\1\end{bmatrix}$	$-G-H, \begin{bmatrix}G\\H\end{bmatrix}$	SS, $G + H > 0$ SU, $G + H < 0$

4.5.2 Coarse numerical methods and their uses.

In the preceding section we saw that the partial differential equation for diffusion and reaction in a catalyst particle could be reduced to

$$\nabla^2 \xi + R(\xi) = 0 \text{ in } \Omega$$

$$\xi = 0 \text{ on } \partial\Omega. \tag{4.66 bis}$$

Let us suppose that Ω is either an infinite slab, an infinite cylinder or a sphere so that the equation becomes

$$\frac{1}{\rho^q}\frac{d}{d\rho}\left[\rho^q \frac{d\xi}{d\rho}\right] + R(\xi) = 0, \tag{4.81}$$

$$\xi = 0, \qquad \rho = 1, \tag{4.82}$$

with $q = 0$, 1 or 2 and ρ is the fractional radius or distance from the central plane. For reasons that we will not go into here (cf. [14]), the form of $R(\xi)$ is:

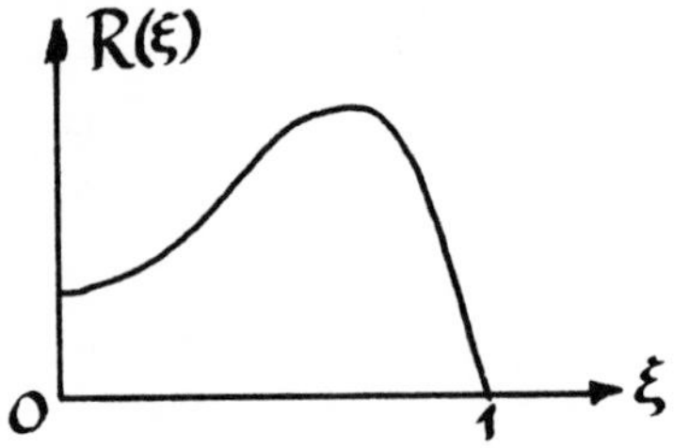

This is clearly a very nonlinear function but is positive so that the solution is a convex function of ρ. In the case of the slab we can solve the equation by quadrature but we wish to follow a route opened by Villadsen and Stewart [188] and use a one point collocation method. For fuller details of this problem Finlayson's monograph [66] or Villadsen and Michelsen's book [261] should be consulted.

Granted that the solution is a convex function of ρ and zero at $\rho=1$, a crude approximation to it would be $\xi(\rho) = \alpha(1-\rho^2)$ and its Laplacian is

$$-2(q+1)\alpha = -2(q+1)\xi(\rho)/(1-\rho^2).$$

Hence if α were chosen to satisfy the equation at a given point, say $\rho=\alpha$, the value of $\xi(\sigma)=\zeta$ would satisfy

$$\frac{2(q+1)}{1-\sigma^2}\zeta = R(\zeta) \tag{4.83}$$

There is a full-blooded theory for the choice of collocation points such as σ, but we will give only the simplest motivation. If $q=2$ and $R(\xi) = \phi^2(1-\xi)$ is linear then eqn. (4.81) can be solved to give:

$$\xi(\rho) = 1 - \frac{\sinh\phi\rho}{\rho\sinh\phi}$$

If the same $R(\xi)$ is used in eqn. (4.83) with $q=2$,

$$\xi(\rho) = \frac{\phi^2(1-\rho^2)}{6+\phi^2(1-\sigma^2)}$$

These expressions should agree as far as possible and we find that they do so for small ϕ if $\sigma^2=3/7$. A similar comparison for other cases gives $\sigma^2=(q+1)/(q+5)$, giving the equation

$$\frac{(q+1)(q+5)}{2}\zeta = R(\zeta) \tag{4.84}$$

for $\zeta=\xi(\sigma)$ and $\alpha=\zeta/(1-\sigma^2) = (q+5)\zeta/4$.

Returning now to the nonlinear form of $R(\xi)$ we have an immediate graphical construction for ζ by drawing a line of the appropriate slope from the origin. When we do this we see there will be

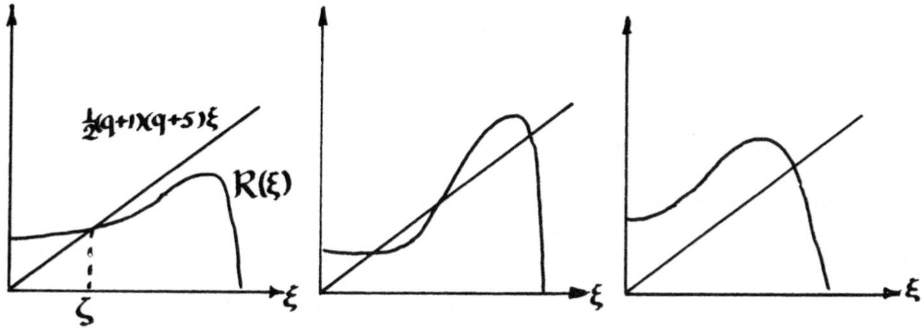

cases where the line will intersect more than once with the curve $R(\xi)$. We are thus warned of the possibility of multiple solutions of the original equation even though we acknowledge that the approximation is crude to a degree. But we also note that if $R(\xi)/\xi$ is monotonic decreasing as ξ increases from 0 to 1, then there can never be more than one intersection. This suggests a criterion for uniqueness that might be usable in the original partial differential equation. Suppose ξ_1 and ξ_2 are two distinct solutions of $\nabla^2\xi + R(\xi) = 0$. By Green's theorems and the fact that $\xi_1=\xi_2=0$ and Ω

$$\iiint_\Omega \{(\xi_1\nabla^2\xi_2 - \xi_2\nabla^2\xi_1)dV = 0$$

Hence substituting from the equations for the Laplacians,

$$0 = \iiint_\Omega \{\xi_1 R(\xi_2) - \xi_2 R(\xi_1)\}\ dV$$

$$= \iiint_\Omega \left[\frac{R(\xi_2)}{\xi_2} - \frac{R(\xi_1)}{\xi_1}\right] \xi_1\xi_2 dV.$$

Now it can be shown (see [105]) that there is a maximal solution and if we choose ξ_1 to be this $\xi_2 \leq \xi_1$ everywhere. If then $R(\xi)/\xi$ is monotonic the integral is positive and can only be zero if $\xi_2=\xi_1$. The monotonicity of $R(\xi)/\xi$ is thus a very general sufficient condition for uniqueness (cf. [122]).

4.5.3 The interaction of easier and more difficult problems.

In this final example we want to explore further aspects of the craft of seeing what can be learned about the model by getting a qualitative feel for the solution before any actual calculation is done. In particular the interplay of the various versions of a model will be emphasized and we shall see how appeal to an apparently more difficult problem can sometimes illuminate a simpler one. The physical system S is a vertical counter-current contacting column through which a gas flows upwards while a solid falls downwards through the gas flow and is removed at the bottom. A substance is brought in with the gas stream where its concentration at height x and time t is $c(x,t)$. It can be adsorbed on the solid where its concentration is denoted by $n(x,t)$. Both these concentrations are the amounts of A per unit volume of the phase and ε and $(1-\varepsilon)$ are the volume fractions of the two phases. Once adsorbed on the solid A can either be desorbed back into the gas or react to form a product B which is instantaneously desorbed. Because of this instantaneous desorption we have no need to consider an equation for the concentration of B simultaneously with those for the concentrations of A, but can determine this after the solution for A. The scheme is $A \rightleftharpoons A^* \rightarrow B$, where A^* denotes the absorbed form of A. The figure shows the notation and general scheme. We shall not dwell on the hypotheses nor on the reduction of the equations to dimensionless form since we are interested in the subsequent treatment of the equations. Suffice it to say that the rate of adsorption is $k_a(N-n)c$ where N is the saturation value of n, and the rates of desorption and reaction are $k_d n$ and $k_r n$ respectively. Then the usual balances give

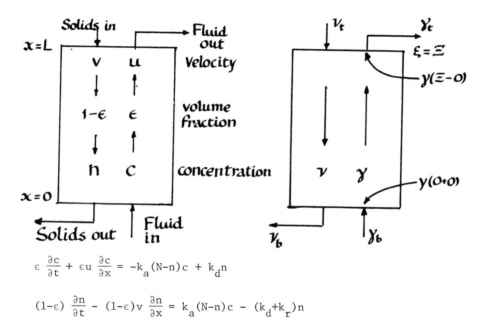

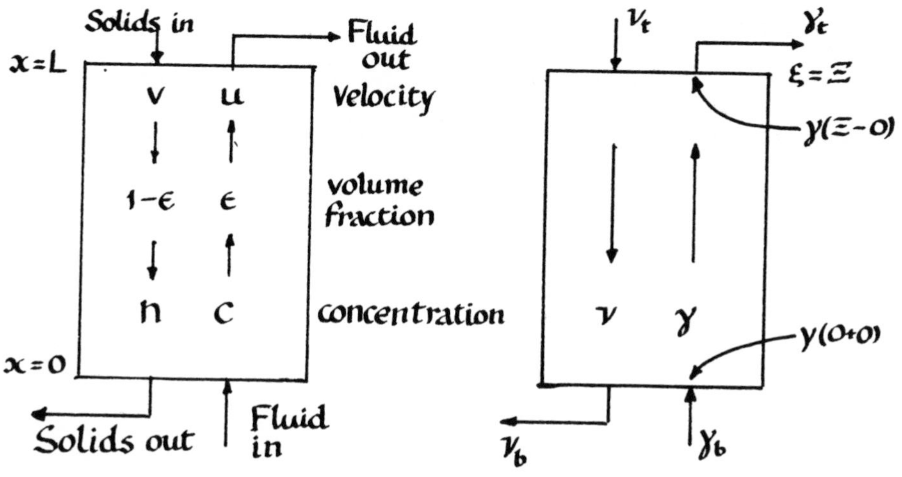

$$\varepsilon \frac{\partial c}{\partial t} + \varepsilon u \frac{\partial c}{\partial x} = -k_a(N-n)c + k_d n$$

$$(1-\varepsilon) \frac{\partial n}{\partial t} - (1-\varepsilon)v \frac{\partial n}{\partial x} = k_a(N-n)c - (k_d+k_r)n$$

These equations can be added to give

$$\frac{\partial}{\partial t} \{\varepsilon c+(1-\varepsilon)n\} + \frac{\partial}{\partial x} \{\varepsilon uc-(1-\varepsilon)vn\} = -k_r n$$

which would be a convenient form if n were known as a function of c.

They will be rendered dimensionless by the following transformations.

$$\gamma = Kc = k_a c/k_d, \quad \nu = n/N$$

$$\xi = k_r x/\varepsilon u, \quad \Xi = k_r L/\varepsilon u, \quad \tau = k_r t/\varepsilon$$

$$\lambda = Nk_a/k_r, \quad \mu = Nk_a/k_d, \quad \alpha = \mu(1-\varepsilon)/\varepsilon, \quad \sigma = \alpha v/u$$

There is no need to swell on this reduction except to draw attention to the meaning of α and σ. $\mu = NK$ is the ratio of adsorbed to fluid phase concentrations in the limit of dilute equilibrium and, in fact, is an upper bound of this ratio. In α this is multiplied by the volume ratio of the two phases and so is the ratio of the amounts that can be held in each. In σ this is further multiplied by the ratio of the velocities and so is the ratio of the fluxes. This interpretation should already suggest that the case of $\sigma>1$ may have some features that are particularly different from $\sigma<1$ since the carrying capacity of the solid stream is then greater than that of the fluid. The three equations are:

$$\frac{\partial\gamma}{\partial\tau} + \frac{\partial\gamma}{\partial\xi} = -\lambda\{(1-\nu)\gamma-\nu\} \tag{4.85}$$

$$\alpha\frac{\partial\nu}{\partial\tau} - \sigma\frac{\partial\nu}{\partial\xi} = \lambda\{(1-\nu)\gamma-\nu\} - \mu\nu \tag{4.86}$$

and

$$\frac{\partial}{\partial\tau}\{\gamma+\alpha\nu\} + \frac{\partial}{\partial\xi}\{\gamma - \sigma\nu\} + \mu\nu = 0 \tag{4.87}$$

We have four models according as we consider the rate of adsorption to be finite or infinite (i.e. λ finite or $\lambda\to\infty$) and according as we consider the transient or steady state. If $\lambda\to\infty$ the only way in which the right hand side of eqn. (4.85) can remain finite is for the equilibrium relation $(1-\nu)\gamma-\nu = 0$ to obtain. Thus

$$\nu = \gamma/(1+\gamma)$$

and, substituting in eqn. (4.87)

$$\left\{1 + \frac{\alpha}{(1+\gamma)^2}\right\} \frac{\partial\gamma}{\partial\tau} + \left\{1 - \frac{\sigma}{(1+\gamma)^2}\right\} \frac{\partial\gamma}{\partial\xi} + \frac{\mu\gamma}{1+\gamma} = 0 \tag{4.88}$$

In the steady state

$$\frac{d\gamma}{d\xi} = -\lambda\{(1-\nu)\gamma-\nu\} \tag{4.89}$$

$$\frac{d\nu}{d\xi} = -\lambda\{(1-\nu)\gamma-\nu\} + \mu\nu \tag{4.90}$$

or

$$\left\{1 - \frac{\sigma}{(1+\gamma)^2}\right\} \frac{d\gamma}{d\xi} + \frac{\mu\gamma}{1+\gamma} = 0 \tag{4.91}$$

The four models are:

Model	Non-equilibrium λ finite	Equilibrium $\lambda\to\infty$
Transient	4.85, 86, 92-95 $:\Sigma$	4.88, 92, 94-5 $:\Sigma_e$
Steady-state	4.89, 90, 94-5 $:\Sigma_s$	4.91, 94-5 $:\Sigma_{es}$

For simplicity we shall consider only constant inlet and boundary conditions. The initial conditions are

$$\gamma(\xi,0) = \gamma_o \tag{4.92}$$

$$\nu(\xi,0) = \nu_o. \tag{4.93}$$

The inlet conditions specify γ at the bottom and just outside the column, whereas ν is specified just above the top. We write these as

$$\gamma(0-0,\tau) = \gamma_b \tag{4.94}$$

$$\nu(\Xi+0,\tau) = \nu_t. \tag{4.95}$$

This is physically correct but we should sense that there may be a problem since Σ_{es} would appear to be overdetermined. The resolution of this difficulty will appear.

We are interested in how these models can be used to illuminate one another and how the scope of the solutions can be understood and their general form obtained without actually computing anything in detail. Obviously Σ is the most difficult model to crack and though there are well-known methods of treating such equations it would be foolhardy to do so without preliminary consideration. We already have the limiting case $\lambda\to\infty$ under consideration and the solution for large λ will presumably be a slight blurring of this limiting case. The limit $\lambda\to\infty$ is not of the same interest for in this case the adsorption equilibrium is so slow that the two streams do not exchange matter at all. Σ_s and Σ_{es} should be obtainable by letting time run out to infinity, but let us start with the simplest model Σ_{es}.

The equation for Σ_{es} is separable and gives

$$\mu\xi = \int_{\gamma(\xi)}^{\gamma(0)} \left\{ \frac{1+\gamma}{\gamma} - \frac{\sigma}{\gamma(1+\gamma)} \right\} d\gamma \tag{4.96}$$

This can be easily integrated but, for the moment, let us examine it without explicitly integrating, since a similar, but not so easily evaluated integral, might arise in a more complex problem. If $\sigma<1$ the integrand is always positive and, since the integrand behaves like $(1-\sigma)/\gamma$ for small values of γ, it is clear that $\xi\to\infty$ as $\gamma(\xi)\to 0$ and a solution can be found for any value of Ξ. In fact all that is needed is to take a segment of

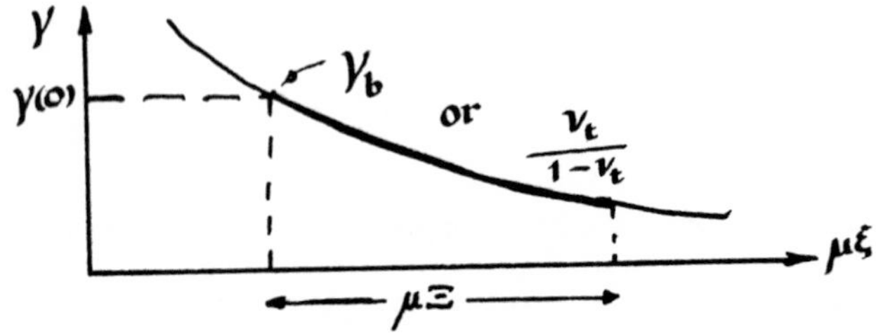

length $\mu\Xi$ of the curve. If we set $\gamma(0)=\gamma_b$ this defines the segment. But we could equally well suppose that the value of $\gamma(\Xi)$ is equilibrium with ν_t and this would fix the segment by fixing the value of γ at the top. Having both conditions is physically sensible, but, mathematically, it is too much and, except in very special circumstances, is going to be contradictory. We might reconcile the physical necessity of specifying both γ_b and γ_t with the mathematical impossibility of generally satisfying both $\gamma(0)=\gamma_b$, $\gamma(\Xi)=\nu_t/(1-\nu_t)$, by saying that there must be a discontinuity at one end or the other. Thus the solid as soon as it contacts the gas at $\xi=\Xi-0$ might (thanks to the fact that λ is infinite) instantaneously take up the equilibrium value $\nu(\Xi)=\gamma(\Xi)/\{1+\gamma(\Xi)\}$. In order for there to be no continuous build-up of matter at the top the concentration γ must also have a discontinuity from $\gamma(\Xi)=\gamma(\Xi-0)$ to $\gamma_t=\gamma(\Xi+0)$. A balance round the plane $\xi=\Xi$ shows that the condition for no accumulation in the plane is

$$\gamma(\Xi) - \gamma_t = \sigma\{\nu(\Xi)-\nu_t\}$$

Thus, if the discontinuity is at the top, the concentration in the emerging fluid is not $\gamma(\Xi)$, but

$$\gamma_t = \gamma(\Xi) - \sigma\left\{\frac{\gamma(\Xi)}{1+\gamma(\Xi)} - \nu_t\right\}. \tag{4.97}$$

But why should this happen at the top? From the mathematics it could equally well happen at the bottom or, for that matter, at both ends. We could argue physically that the case $\sigma<1$ must be continuous with $\sigma=0$ and that in this case the bed of solid is fixed and only the condition at the bottom can be insisted upon. But to see this mathematically we have to look at one or other of the harder problems Σ_e or Σ_s to get the needed insight into the allegedly simpler Σ_{es}. Let us pick Σ_e leaving the reader to look into Σ_s, since more will be said about this later.

Σ_e is a single first-order quasilinear partial differential equation which can be readily solved by the method of characteristics. In fact the characteristic equations of eqn. (4.88) are

$$\frac{d\tau}{ds} = 1 + \frac{\alpha}{(1+\gamma)^2} \tag{4.98}$$

$$\frac{d\xi}{ds} = 1 - \frac{\sigma}{(1+\gamma)^2} \tag{4.99}$$

$$\frac{d\gamma}{ds} = - \frac{\mu\gamma}{1+\gamma} \tag{4.100}$$

Again these are easy enough to integrate explicitly but, having in mind that we want to adopt tactics that would work with more complicated equations, we will lightly avoid this. We first notice that γ decreases monotonically along a characteristic so that it might be taken as the parametric variable along the characteristic. Clearly τ always increases along the characteristic and, since $\sigma<1$, so does ξ. Thus the characteristics always have a positive slope. In the ξ,τ plane we are interested in the strip $\tau>0$, $0 < \xi < \Xi$ and this can be covered by characteristics emanating from the initial interval $\tau=0$, $0 < \xi < \Xi$ where $\gamma=\gamma_o$ and the inlet at the bottom $\xi=0$, $\tau<0$ where $\gamma=\gamma_b$. The characteristics emanating from points on

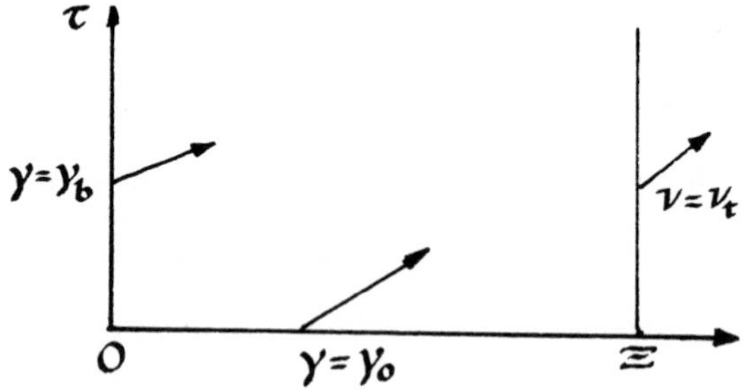

the top, $\xi=\Xi$, are directed outwith the region of interest so that conditions specified at the top can have no influence there. The discontinuity described by eqn. (4.97) must therefore be present at the top of the column. This is entirely in keeping with our feelings about the meaning of $\sigma<1$, for the fact that the carrying capacity of the solids is less than that of the fluid means that the fluid stream can "blow out" the influence of the solids. Thus ν_t has no influence at all on the solution inside the reactor, but it does have an influence, through eqn. (4.97) on γ_t, the product of the reactor. This resolution satisfies both our mathematical and physical expectations.

To complete the case $\sigma<1$ let us note that the slope of characteristic

$$\left[\frac{d\tau}{d\xi}\right]_c = \frac{(1+\gamma)^2+\alpha}{(1+\gamma)^2-\sigma} \tag{4.101}$$

is positive and increases as γ decreases going from near 1 when γ is large to $(1+\alpha)/(1-\sigma)$ as $\gamma\to 0$ (see section (a) of figures below). If $\gamma_b<\gamma_o$

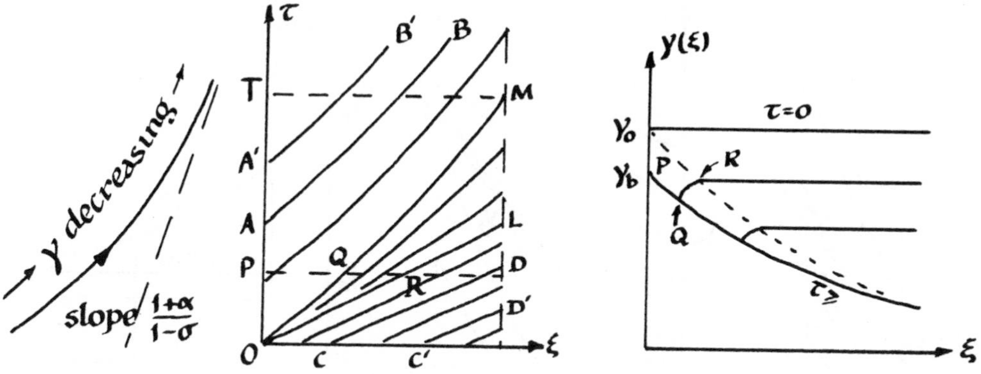

(a) (b) (c)

the characteristics emanating from $\xi=0$ all have greater slope than those that emanate from $\tau=0$. There is thus no tendency for the two sets of characteristics to overlap one another; the one set is vertically parallel like AB and A'B' and the other horizontally parallel like CD and C'D'. Moreover at the origin all concentrations between γ_o and γ_b may be thought to be present. The characteristics corresponding to these therefore fan out and fill in the region LOM. Taking sections at constant values of τ we see that the profile of the steady state is established in the length of reactor up to the characteristic OM then there is a transition region between OM and OL beyond which the initial constant state is decaying and being pushed out of the reactor. The final profile (PS in Section C of the figure) is established in the time T).

On the other hand if $\gamma_o<\gamma_b$ the characteristics from $\xi=0$ have a smaller slope than those from $\tau=0$ and will overlap one another and all the characteristics emanating from the origin (section (a) of figure below). This is intolerable as it would imply that three different concentrations were present at one point. The resolution of this is to introduce a shock or discontinuity. The speed of a shock, w, is such that its movement during

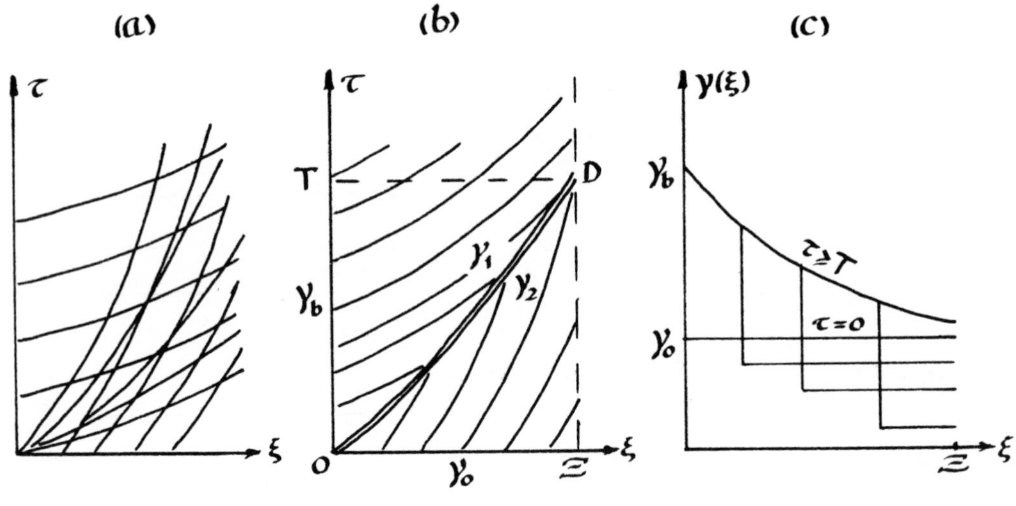

a time interval δt, namely $w\delta t$, accounts for the net access of material. Thus if c_1,n_1 are the values of the concentrations just below and c_2,n_2 those just above the discontinuity

$$w\delta t[\varepsilon(c_1-c_2)+(1-\varepsilon)(n_1-n_2)]\delta t$$

$$= [u\varepsilon(c_1-c_2)-v(1-\varepsilon)(n_1-n_2)]\delta t$$

or

$$\left[\frac{d\xi}{d\tau}\right]_d = \frac{w}{u} = \frac{[\gamma] - \sigma[\nu]}{[\gamma] + \alpha[\nu]}$$

where $[\gamma] = \gamma_1 - \gamma_2$ and $[\nu] = \nu_1 - \nu_2 = [\gamma]/(1+\gamma_1)(1+\gamma_2)$.

Thus the slope of a shock line in the ξ,τ-plane is

$$\zeta_{12} = \left[\frac{d\tau}{d\xi}\right]_d = \frac{(1+\gamma_1)(1+\gamma_2)+\alpha}{(1+\gamma_1)(1+\gamma_2)-\sigma} \tag{4.102}$$

This equation is confirmed by the fact that when the discontinuity vanishes, i.e. $\gamma_1 = \gamma_2 = \gamma$, we have the slope of the characteristic. We also note that ζ_{12} is certainly positive as long as $\sigma<1$. If ξ_s, τ_s is a point on the shock line with γ_1 behind and γ_2 before we can calculate the path of the shock as follows. γ_1 must correspond to a distance ξ_s along a characteristic that emanates from $\xi=0$ where $\gamma=\gamma_b$. Thus, dividing eqn. (4.98) by eqn. (4.100) and separating variables,

$$\mu\xi_s = \int_{\gamma_1}^{\gamma_b} \left[\frac{1+\gamma}{\gamma} - \frac{\sigma}{\gamma(1+\gamma)}\right] d\gamma \tag{4.103}$$

Similarly, γ_2 is the value of γ on a characteristic emanating from $\tau=0$ where $\gamma=\gamma$; thus from eqns. (4.99) and (4.100)

$$\mu\tau_s = \int_{\gamma_2}^{\gamma_0} \left[\frac{1+\gamma}{\gamma} + \frac{\alpha}{\gamma(1+\gamma)}\right] d\gamma \tag{4.104}$$

Moreover $d\tau_s/d\xi_s = \zeta_{12}$ is given by eqn. (4.102) so that these equations, 4.102, 103 and 104, provide a way of calculating the path of the shock. We are not concerned here with the question of how these equations can be solved, but the solution must clearly give a path such as OD above. The final steady state solution is fully developed behind the shock while the initial state decays and is pushed out in front (i.e. above) it. Again the steady state is established in a finite, calculable time T.

Let us now consider the case $\sigma>1$ and return at first to the steady state equation of Σ_{es}. The solution is again given by eqn. (4.96) for sufficiently small values of ξ. But we notice that when γ drops to $(\sigma^{1/2}-1)$ the integrand vanishes and subsequently becomes negative. This means that as $\gamma(\xi)$ continues to fall the value of ξ decreases instead of increasing, giving the solution shown in the figure. But this is physically impossible since it gives two values of γ for some ξ and cannot give a solution if $\Xi>\xi'$. This suggests that there must be discontinuities in the solution and to see where these arise we turn this time to Σ_s.

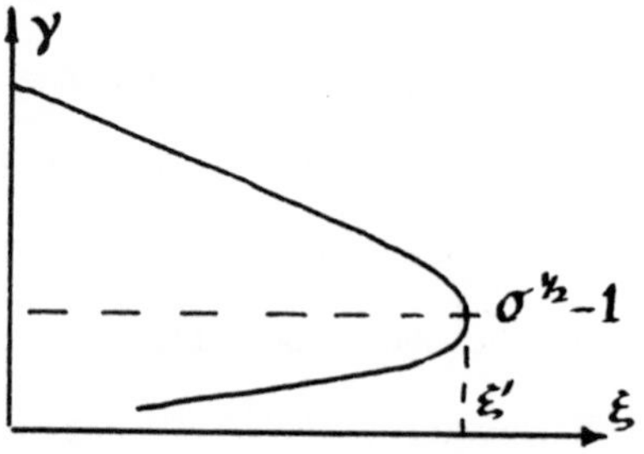

Σ_s is a pair of ordinary differential equations, (4.89) and (4.90), whose right hand sides are functions only of γ and ν. We can combine them into a single differential equation which will give the relation between ν and γ at any point by dividing the one by the other

$$\frac{d\nu}{d\gamma} = \frac{1}{\sigma} - \frac{\mu}{\lambda\sigma}\,\frac{\nu}{(1-\nu)\gamma-\nu} \tag{4.105}$$

The only critical point is the origin, which is a saddle-point having an entering trajectory of slope η_- and departing trajectory of slope η_+ where

$$\eta_\pm = \frac{1}{2\sigma}\left[1 + \sigma + (\mu/\lambda) \pm \sqrt{\{1+\sigma+(\mu/\lambda)\}^2-4\sigma}.\right]$$

In fact the isoclines $d\nu/d\gamma=\eta$, are

$$\nu = \frac{(\lambda/\mu)(1-\sigma\eta)\gamma}{1 + (\lambda/\mu)(1-\sigma\eta)(1+\gamma)} = \frac{\beta\gamma}{1+\beta\gamma} \tag{4.106}$$

where

$$\beta = \frac{\lambda(1-\sigma\eta)}{\mu+\lambda(1-\sigma\eta)} \quad .$$

In particular there are the two loci of horizontal ($\eta=0$) and vertical ($\eta=\infty$) directions and we note that $\eta_+ > \beta_\infty > \beta_o > \eta_-$. Thus the phase plane looks as below where OA and OB are the trajectories through the saddle point O and OC and OD the isoclines of verticality and horizontality.

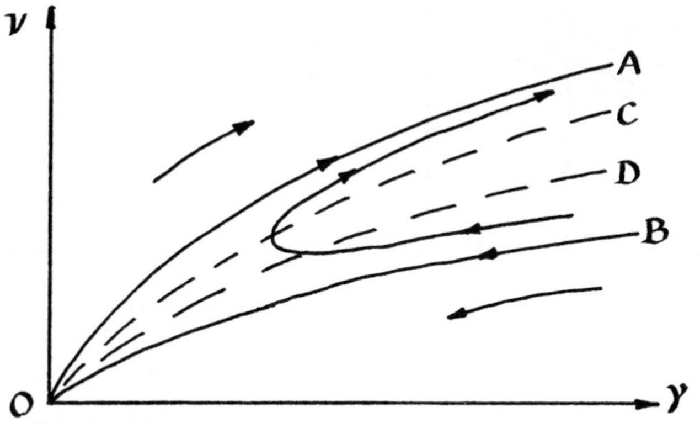

To solve eqn. 4.105 subject to the conditions $\gamma(0)=\gamma_b, \nu(\Xi) = \nu_t$ we have to find a path in the phase plane between the vertical line $\gamma=\gamma_b$ and the horizontal $\nu=\nu_t$. Not any such trajectory will do but if the reactor is of length Ξ we must find the path for which eqn. (4.89) gives

$$\lambda \, \Xi = \int_{\gamma_b}^{\nu_t} \frac{d\gamma}{(1-\nu)\gamma-\nu} \quad .$$

This integral requires a bit of interpretation when the path crosses $\nu=\gamma/(1+\gamma)$ since the integrand there becomes infinite. If $ds^2 = d\gamma^2 + d\nu^2$ is the path length along the trajectory and $f = f(\gamma,\nu) = (1-\nu)\gamma-\nu$, then the integral could be written

$$\lambda \, \Xi = \int \frac{\sigma ds}{\{\sigma^2 f^2 + (f-\mu\nu)^2\}^{1/2}} \, , \tag{4.107}$$

a form which is unexceptionable. For example, if the point (γ_b,ν_t) lies between the arms OA and OB there are a number of possible paths: P itself of length zero; QR corresponding to a short length; ST to a longer reactor and UV to a very long one; the two segments XO and OY

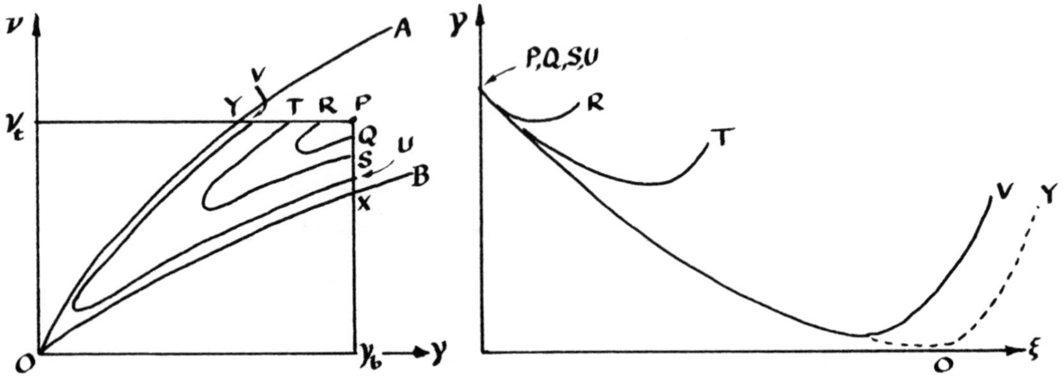

of OB and OA would give an infinite reactor since it takes an infinite length to get into and out of the origin. If the point (γ_b, ν_t) lies beneath OB the situation is somewhat different as shown below. Again

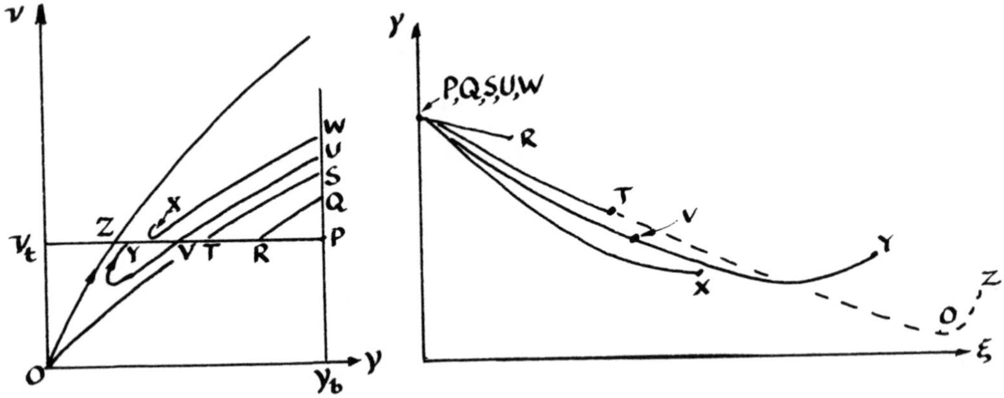

there is the path of zero length P, a short path QR and longer one s, ST and UV. These are monotonic until WX, but to fit a reactor longer than the Ξ corresponding to the path WX we must go back to a curve such as UXY which follows an earlier solution UV but dips below $\nu = \nu_t$ and comes up again. STOZ is the ultimate solution of this kind since it takes an infinite length to get into and out of O. (Note that paths such as ZA do not generate solutions since they go from the horizontal $\nu=\nu_t$ to the vertical $\gamma=\gamma_t$ and not vice-versa).

How does the situation look in the limit $\lambda\to\infty$? This will take us back to Σ_{es} where we foresaw certain difficulties in getting any solution at all if $\Xi>\xi'$. We have no space to classify exhaustively all the possibilities (this has been done by Aris and Viswanathan [18]) so will take only one case. As $\lambda\to\infty$ all the isoclines in the figure on p. 97 close down on the one curve $\nu=\gamma/(1+\gamma)$ and this curve, Γ, is a solution of the equation. This is not surprising for $\lambda\to\infty$ means that adsorption equilibrium is instantaneous so that we should expect the equilibrium relationship to obtain. At any point not on Γ the limit $\lambda\to\infty$ in eqn. 4.105 gives $d\nu/d\gamma = 1/\sigma$, i.e. the solution is a straight line. Such a solution corresponds to a discontinuity for with λ infinite any path integral such as eqn. 4.107 shows that a segment of such a line is traversed in zero length. If $\sigma>1$ there is a point, C, on the curve Γ of slope $1/\sigma$, in fact $\gamma=\sigma^{1/2}-1$ at this point. Below it the curve is traversed upwards and above the trajectory is downwards. The phase plane is:- where all the straight line segments all correspond to discontinuities.

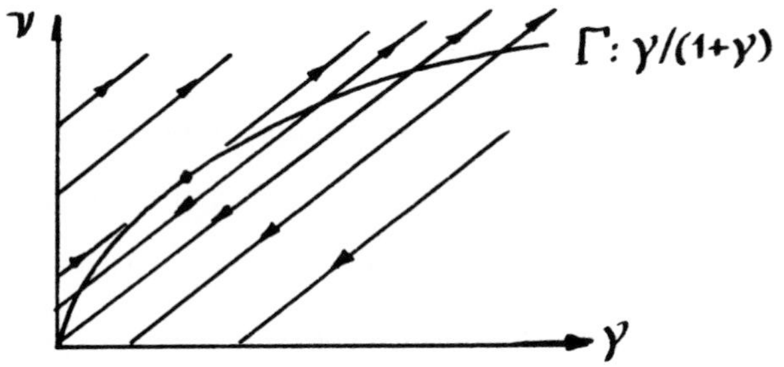

Consider a pair of boundary conditions γ_b, ν_t that give a point close under the upper part of Γ as shown (in fact $\gamma_b>\sigma-1$, $1-\sigma^{-1/2}<\nu_t<\gamma_b/(1+\gamma_b)$ as the longer analysis of Aris and Viswanathan shows). The possible ways of getting from the vertical $\gamma=\gamma_b$ to the horizontal $\nu=\nu_t$ are shown numbered in order of increasing reactor length. 1 is trivial corresponding to a reactor of zero length. In 2 the reactor is so short that the concentration does not fall very much in the reactor, it is continuous at the bottom but has a discontinuity at the top. 3 is the segment of Γ between $\gamma=\gamma_b$ and $\nu=\nu_t$ and corresponds to the only length Ξ for which there is a continuous solution in the closed interval $0 < \xi < \Xi$. For a slightly longer reactor, 4, we can go below the horizontal and come back to it with a discontinuity at the exit and this can be done until we reach the point C. This point, where $\gamma(\Xi)=\sigma^{1/2}-1$, corresponds to a reactor of length ξ' as in the figure of p. 96. How do we get a one-valued

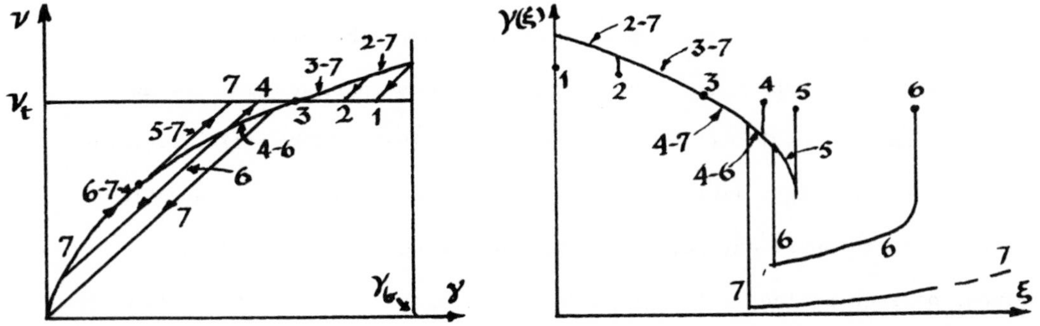

solution if $\Xi>\xi$? The only way is to introduce a discontinuity in the middle of the solution, following the curve Γ downwards to a point short of C, then taking a chord, such as 6, down to a point below C, following Γ back up to C and taking a jump to $\nu=\nu_t$. For all such solutions $\gamma(\Xi) = \sigma^{1/2}-1$. This type of solution can be fitted into any reactor that is long ($\Xi>\xi$), for a path such as 7 can be made very long since ξ moves rapidly with γ in the neighbourhood of the origin. Without doing any real calculation we have seen how the solution must lie.

Finally we may ask how this links up with the transient solution of Σ_e. Returning to the characteristic equations 4.98, 99 and 100 and taking decreasing γ as the parameter along the path of the characteristic we see that the solution of

$$-\mu \frac{d\xi}{d\gamma} = \frac{1+\gamma}{\gamma} - \frac{\sigma}{\gamma(1+\gamma)}$$

and

$$-\mu \frac{d\tau}{d\gamma} = \frac{1+\gamma}{\gamma} + \frac{\alpha}{\gamma(1+\gamma)}$$

is a curve of positive slope for $\gamma>\sigma^{1/2}-1$, which turns back on itself at $\gamma=\sigma^{1/2}-1$ and then has a negative slope ultimately decreasing to $-(\sigma-1)/(\alpha+1)$. We could in fact solve these equations once and for all and cut out a template for any pair of values of α and σ, graduating its edge in γ. It would look like this:

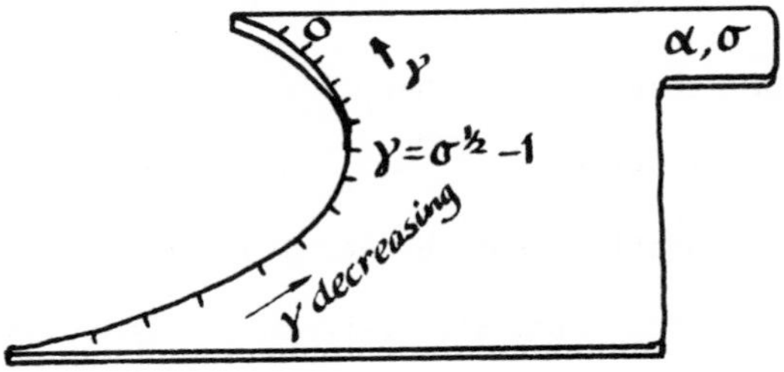

By laying such a template with γ_o at ξ_o, τ_o we can draw the characteristic emanating from such a point. For example, the characteristics emanating from $\xi=0$ can be drawn by placing γ_b on the vertical axis drawing along the template and then moving it upwards to give a sequence of parallel curves. Clearly if the curve is taken beyond $\sigma^{1/2}-1$ these curves can intersect one another so that there is an even greater need for discontinuities than before.

Suppose that γ_b and ν_t are disposed in the way we have just considered, that $\gamma_o=0$ and that the reactor is long ($\Xi>\xi$). Then the characteristics emanating from the axis $\tau=0$ are straight lines of slope $-(\sigma-1)/(\alpha+1)$ which will certainly intersect those emanating from $\xi=0$. The resolution of this is to introduce a shock as we did before. This works well up to A the point where AB, the characteristic though $\Xi-0$ meets

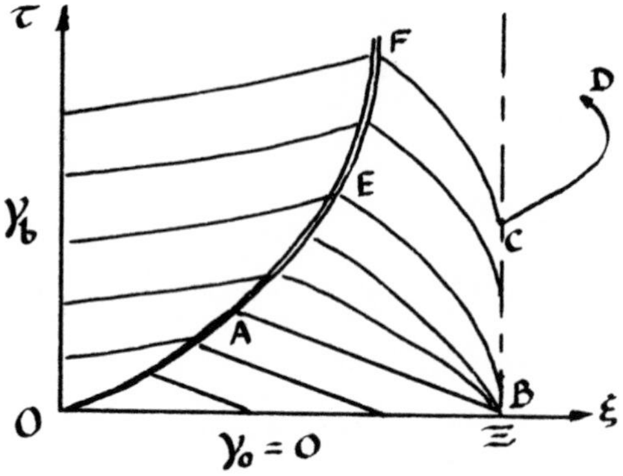

the shock. The characteristic from the vertical line, ξ-Ξ if drawn as starting with $\gamma=\gamma_t=\nu_t/(1-\nu_t)$, the concentration γ in equilibrium with ν_t, would point out of the region like CD, which is not permitted. However we know from the study of Σ_{es} how to resolve this. There is a shock at the top and $\gamma(\Xi)$ is always $\sigma^{1/2}-1$. This allows us to draw all the characteristics emanating from the vertical like BE and CF--they are in fact tangential to the vertical. Between BA and BE there is a fan of characteristics corresponding to emanating concentrations between $\gamma_o=0$ and $\gamma(\Xi)=\sigma^{1/2}-1$. We now have all the characteristics and can start from 0 and find the shock path that will satisfy eqn. 4.102. Cross-sections at a typical sequence of points then show the development of the steady state solution. In the next figure, 0,1,2,3, ∞ are the labels for an increasing sequence of times. 0 is $\tau=0$ when $\gamma=0$; 1 is a time beneath the level of A in the previous figure and there is still a midsection that is unaffected by the inlet conditions either at the bottom or top; 2 is a time between the levels of A and E before the top profile has assumed its final form, 3 is a time above E when all that remains for the discontinuity to move forward to its final position, ∞.

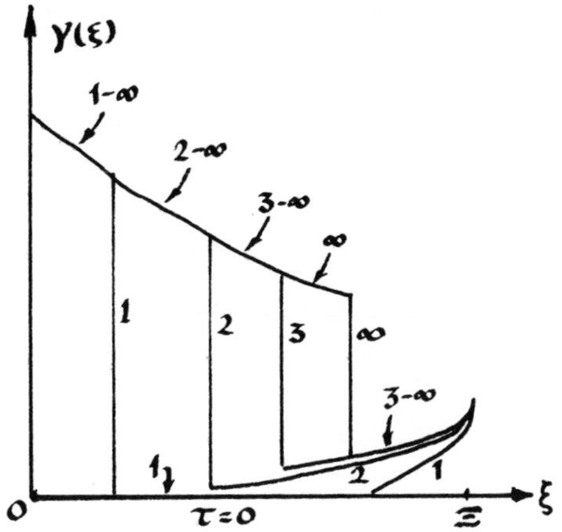

Thus we have built a framework within which the computation of the difficult case Σ can proceed and this framework can be made complete by exploring the whole range of possible dispositions of γ_b, ν_t and γ_o. It is interesting to note that though the internal steady state in Σ_e is only approached asymptotically as the internal shock moves into its final position, the output of the reactor reaches its final value instantaneously since $\gamma(\Xi,\tau)=\sigma^{1/2}-1$ for all τ. This is further discussed in [13].

5 How should a model be evaluated?

'Models are undeniably beautiful, and a man may justly be proud to be seen in their company. But they may have their hidden vices. The question is, after all, not only whether they are good to look at, but whether we can live happily with them.'

A. Kaplan. The Conduct of Inquiry.

5.1 Effective presentation of a model.

A mathematical model and the results that flow from its analysis deserve to be presented effectively. There is nothing meretricious about this, it is merely common-sense. If the model is worth studying and its analysis illuminates the system then it should be presented in such a way that its intrinsic merits or contributed understanding can be quickly grasped and fairly assessed. One of the virtues of a model is that it can be studied more or less comprehensively and with a little care can often ensure that every representative case has been studied. It then becomes a question of presenting it as effectively as possible. The phase plane is a good example of this for by showing a sufficient number of trajectories it allows the eye to visualize all possible solutions. Nor does the number of trajectories have to be large to do this well. The limitation of such phase portraits is that they are difficult to draw in more than two dimensions and one has to make do with a few two-dimensional projections. Certainly coloured holography would allow for four-dimensional presentations, but this is clearly beyond the common reach; stereoscopic diagrams are possible but in their usual form they require a decoupling of the eyes that not everyone can manage (cf. [157]). Nor is the artistic virtuosity of such a journal as Scientific American easily attained, though how effective such graphic art can be is seen in such articles as Zeeman's survey of catastrophe theory [197] (cf. also [255]). Even so the cusp is the highest canonical catastrophe that can be shown in its entirety; the catastrophe set of the swallowtail can be fully pictured, but the butterfly and other canonical forms have to be shown in section. Nevertheless part of the appeal of catastrophe theory is that it permits a more synoptic view of a greater range of behaviour than had been otherwise cultivated. (The literature of catastrophe theory grows apace: cf. [166,177,197,204,205,225,228,234,235, 250-2,255]).

The stirred tank reactor system Σ_6 is an example that can be well presented as topologically equivalent to a cusp catastrophe. Eqns. (4.11) and (4.13) for the steady state with a first order irreversible reaction reduce to eqn. (4.16), namely

$$(v - \bar{v})/\zeta = z(v;\alpha) \tag{5.1}$$

where

$$z(v;\alpha) = \frac{\alpha e^{-1/v}}{1 + \alpha e^{-1/v}} \ . \tag{5.2}$$

When v_s, the steady state temperature is found, both sides of the equation are equal to the steady state reaction rate. So if we let

$$Z = z(v_s;\alpha) = Z(\alpha,\zeta, \bar{v}) \tag{5.3}$$

we see that Z is a function of three parameters: α, the Damköhler number or intensity of reaction; ζ, a combination of heat of reaction and heat removal rate; $\bar{v}$, a mean feed/coolant temperature. If α is held constant the surface $Z(\alpha,\zeta, \bar{v})$ as a function of ζ and $\bar{v}$ takes on the form of a cusp catastrophe as has been known for sixty years. When the cooling capacity is very large $h \to 0$ and $\zeta \to 0$; when the operation is adiabatic $h \to 0$ and $\zeta \to \beta$. The possible modes of intersection are shown in the figure.

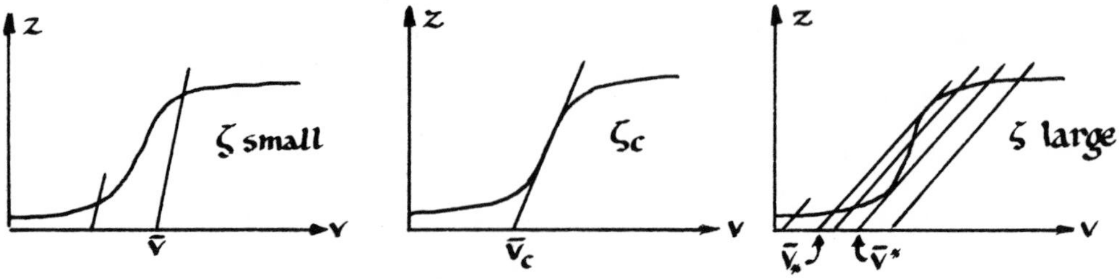

When ζ is small the straight line, which is the left hand side of the equation, is steeper than any part of the z-curve and the intersection is unique whatever the value $\bar{v}$. If ζ has the critical value ζ_c that gives the line the slope of the inflection point of the z-curve, the solution is still unique but the tip of the cusp is attained when $\bar{v}_c$ makes the line go through the inflection point. For $\zeta > \zeta_c$ then, for $\bar{v}_*(\zeta) < \bar{v} < \bar{v}^*(\zeta)$, there are three intersections.

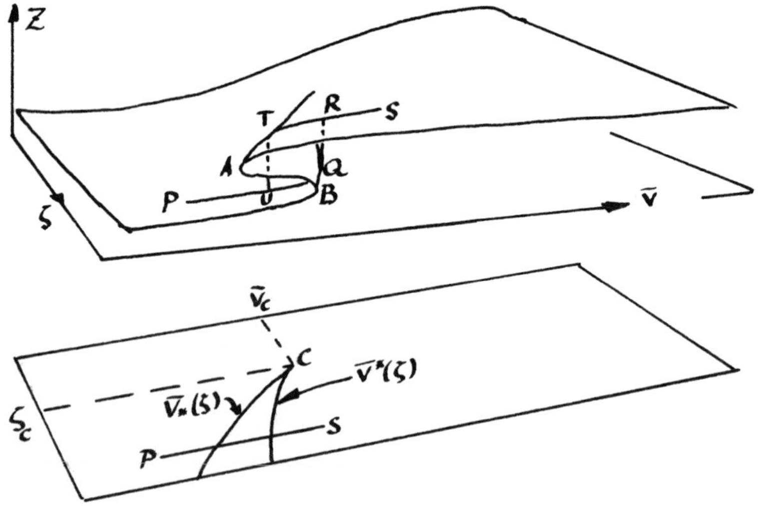

The lower part of the surface (left of CB) corresponds to a poor rate of reaction, where as the upper right (right of CA) represents a much higher reaction rate. The branches CA and CB of the catastrophe set might be called the extinction and ignition catastrophes respectively, for if ζ is constant and $\bar{v}$ is increased from P to S and decreased back again, ignition takes place in a jump from Q to R and extinction in the drop from T to U.

The parameter α has been kept constant in this presentation and as it varies the cusp moves. We shall study this in some detail in the next section and it suffices to remark here that, though the cusp moves, no new features come to light (cf. Regenass and Aris [152]). However, two parameters are affected by the variation of flow rate and as was remarked in Sec. 4.3 a rather different parametrization is appropriate. Using the variables and parameters of eqns. (4.29 and (4.30) we have

$$\Theta\dot{u} = -u + \alpha\Theta(1-u)\exp[v/(1+v/\gamma)] \tag{5.4}$$

$$\Theta\dot{v} = \Theta v_c - (1+\Theta)v + \Theta\alpha\beta(1-u)\exp[v/(1+v/\gamma)] \tag{5.5}$$

This is essentially the same form of equations as that studied by Uppal, Ray and Poore [184] in one of the landmark papers of the subject. Following them, and for reasons we will not digress to here, we let $\gamma\to\infty$ and take $v_c=0$. There are still three parameters and one must be kept constant if we are to draw surfaces in three dimensions. Let β be fixed for the whole figure and plot the steady state reaction rate Z as a function of α and Θ. Such a surface is shown below and it is clearly somewhat different from the simple cusp surface having a Rehoboam-like finger protruding from the top of the wave. The catastrophe set is the hook-shaped region shown on the plane beneath.

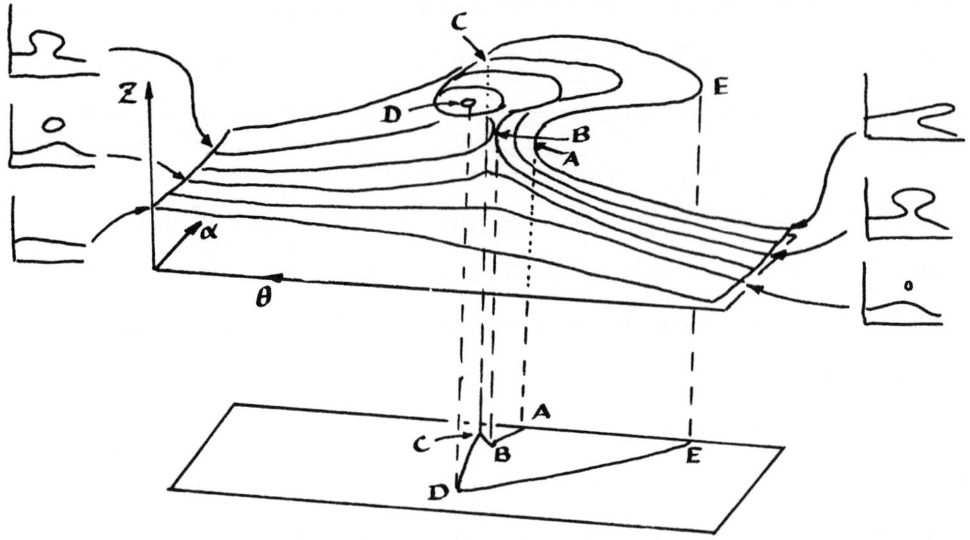

This is one way of presenting a small part of the results of Uppal, Ray and Poore but they have done vastly more than this by applying Hopf bifurcation to eqns. (5.4) and (5.5). By this means they have been able to direct their computations and discover all the possible types of behaviour of the system. These are shown in the next figure. Although the steady state behaviour is sufficiently demarcated by the hook-like catastrophe set (heavy line), within which there are three steady states and outside of which there is uniqueness, the dynamic behaviour is much more complicated. Nine different types of phase portrait can be discerned each corresponding to a different part of the α,Θ plane. Thus if (α,Θ) lies in region B there is a single steady state, but it is unstable and is surrounded by a stable limit cycle. Often a figure such as this will have to be distorted to show the detail, but it then becomes a map to guide the viewer into an accurately drawn figure.

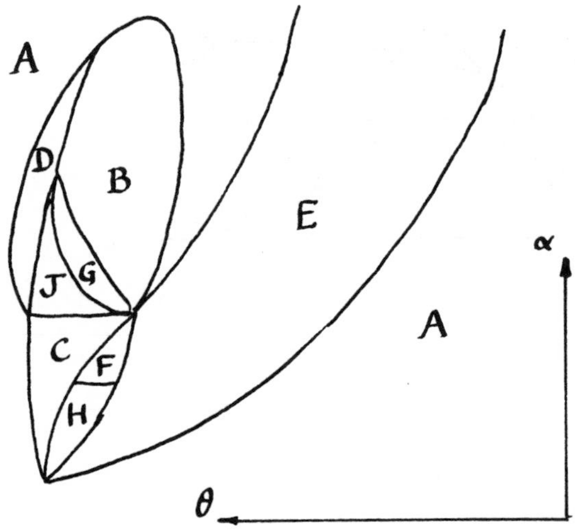

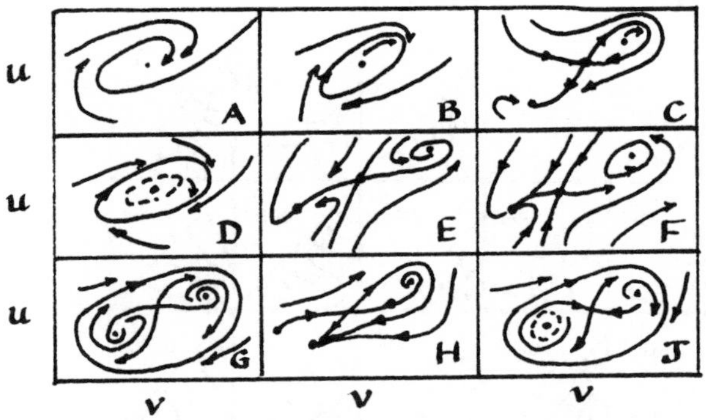

A slightly different form of comprehensive presentation can be given for the system we studied in Sec. 4.5.1. If the growth curves of the two organisms are either of the following forms:

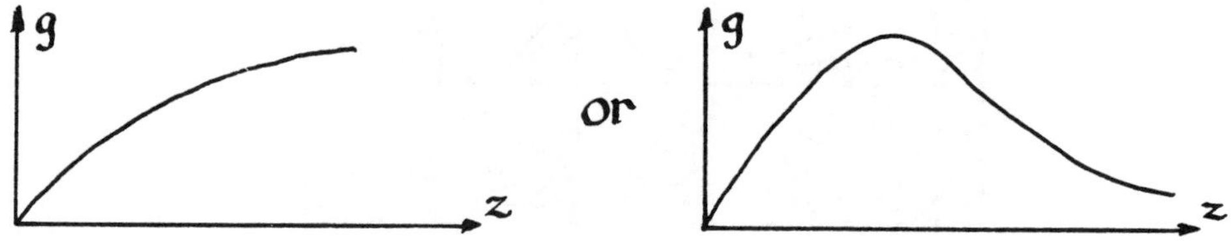

and the label 1 is always used for the curve that is ultimately the higher, i.e. $g_1(z) \geq g_2(z)$ for large z, then we draw both curves in the same plane and put in the point (Z,Θ). This is called 0 and reading from right to

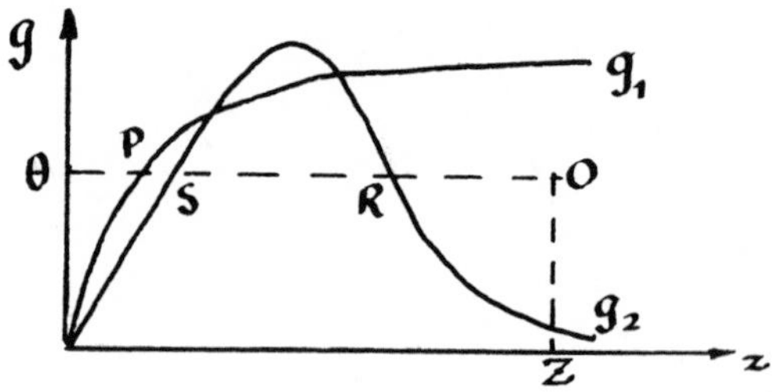

left the intersection with g_1 are first P then Q, if it is needed (it is not in the diagram above) and those with g_2 are first R then S. This gives a word (in the case above ORSP) and the corresponding phase portrait can be found in the gallery below (in the case above #21). This gallery together with the portrait of OPRSQ on p. 81 is complete except for confluence such O $\overline{RS}$

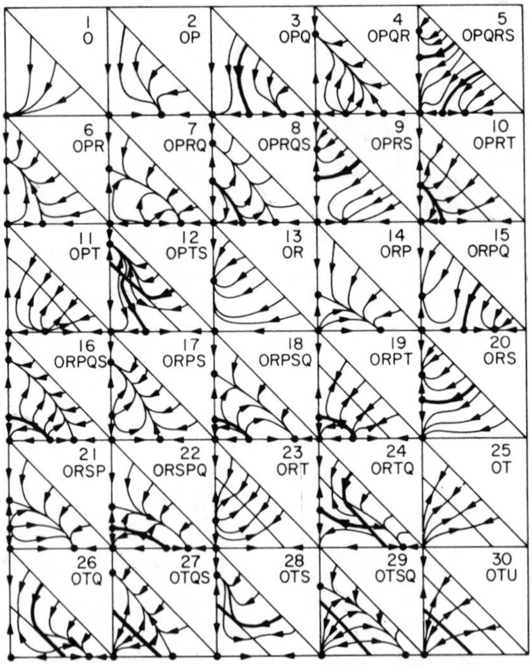

This confluence is not structurally stable for the slightest increase of will make $\overline{RS}$ disappear, while the slightest decrease will separate the confluence and give ORS. We have not been quite consistent in overlooking the structurally unstable however for we have included OT which arises from being exactly at the level of an intersection as part (b) of the next figure shows. We have even included OTU (shown in part (c) of the figure) which is "very unique" since it can only occur if the two intersections are at the same level.

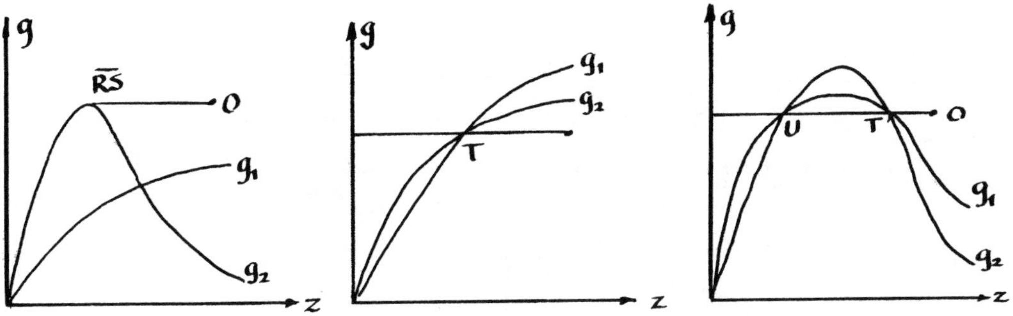

Two further examples of presentation may be given to illustrate the importance of bringing out a characteristic feature in as dramatic a fashion as possible. There are various ways of describing the residence time distribution in a flowing system. $P(t)$, the fraction of particles or molecules in the system at $t=0$ which have left by time t, is a monotone increasing function of t with $P(o)=0$, $P(\infty)=1$. It may be interpreted as the probability of a residence time being less than t. The expected residence time is $\mu = \int tP'(t)dt = \int\{1-P(t)\}dt$ and the variance of residence times $\sigma^2 = \int t\{1-P(t)\}dt-\mu^2$. These are the most obvious parameters and may be related to the position of the 'step' in the P-curve and to its steepness. An alternative description is $p(t)$ the probability density of residence times, i.e. $p(t)dt$ is the probability of a particle having a residence time in the interval $(t,t+dt)$. Clearly $p(t)dt = P(t+dt)-P(t)$, so that $p(t) = P'(t)$, $\mu = \int tp(t)dt$ and $\sigma^2 = \int t^2p(t)dt-\mu^2$. The same information is present but this time the mean is the centre of gravity of a peak and the variance is a measure of its 'spread'. With an approximately Gaussian residence time distribution the one has little advantage over the other.

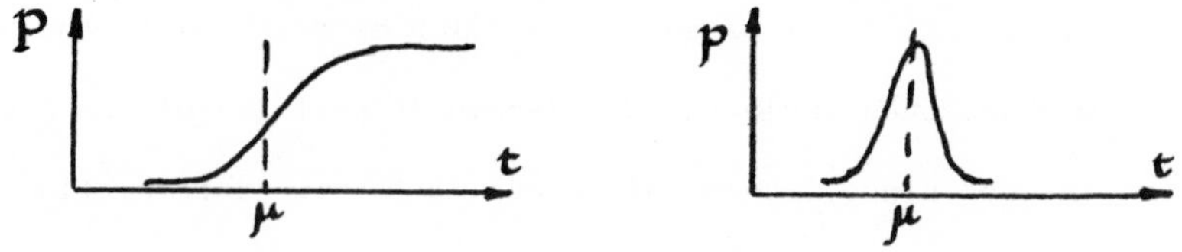

but where there is a rapid bypass the density function shows it up better. For example the system below would give a peak at the bypass residence time in the p-curve but this is rather lost in the P-curve.

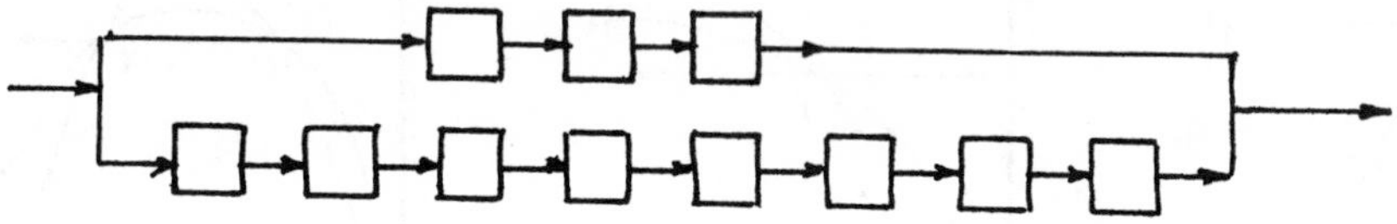

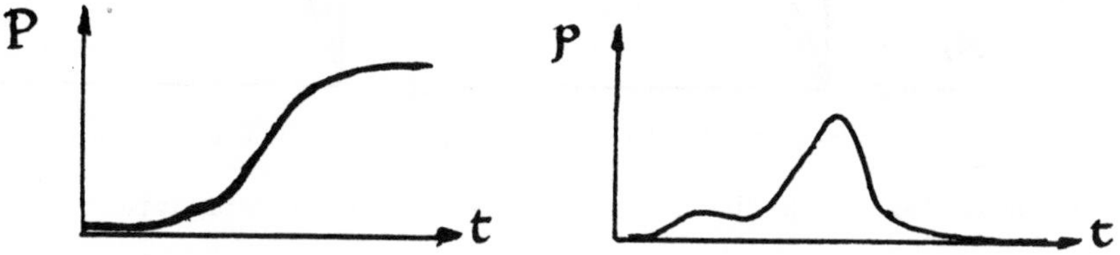

Shinnar has suggested the intensity function or escape probability density as another method of showing up the features of a residence time distribution. This is $\eta(t)dt$ the probability that a molecule of age t will leave the system during the interval $(t,t+dt)$. Since $1-P(t)$ is the fraction of molecules of age t in the system at time t, $p(t)dt = \{1-P(t)\}\eta(t)dt$ or

$$\eta(t) = \frac{p(t)}{1-P(t)} = \frac{d}{dt} \ln\{1-P(t)\}$$

For a single well-mixed compartment

$$p(t) = \frac{1}{\mu} e^{-t/\mu} \quad , \quad P(t) = 1-e^{-t/\mu} \quad , \quad \eta(t) = \frac{1}{\mu} \ .$$

so that the departure of $\eta(t)$ from constancy is an indication of ill-mixedness. Some characteristic shapes are shown below, but for a full discussion of the intelligent use of residence time distributions the papers of Shinnar and his colleagues should be consulted [241-6].

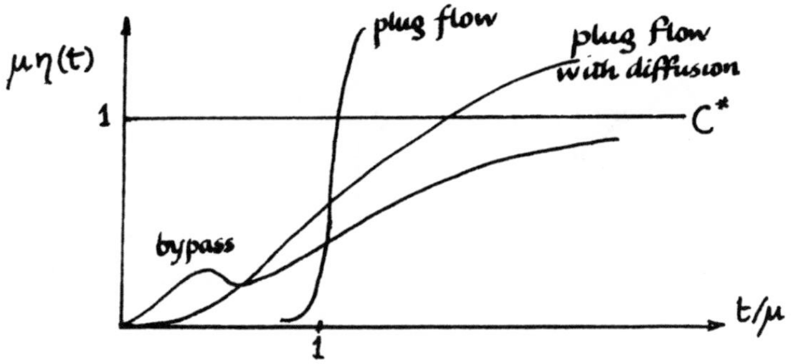

Another situation that lends itself to considered presentation arises when two functions vary in opposite senses, the one increasing and the other decreasing as functions of a controlling variable. Such a situation can obtain when two pollutants behave contrarily with respect to an operating variable

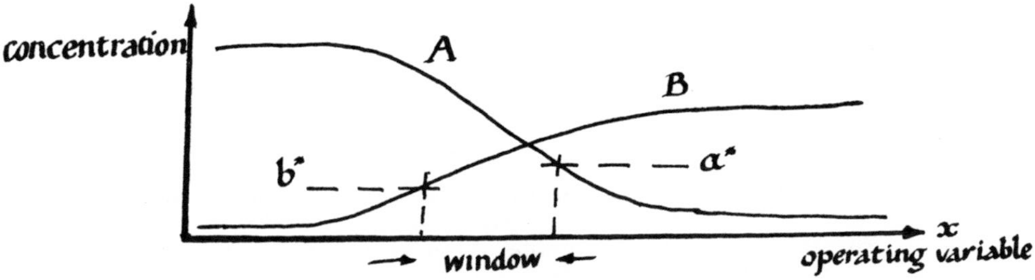

leaving a 'window' of acceptable conditions. If $a(x)$ and $b(x)$ are the concentrations, with $a'(x)<0<b'(x)$, and acceptable levels of A and B are a^* and b^*. If $\alpha(a)$ and $\beta(b)$ are the inverse functions of $a(x)$ and $b(x)$ (i.e. $a(\alpha(a))\equiv a$, $b(\beta(b))\equiv b$), there will be an operating window if $\beta(b^*)<\alpha(a^*)$. However this plot tells us little about the relative behaviour of A and B and it is better to plot the curve $a=a(x), b=b(x)$ in the a,b-plane. The curve may be convex or concave toward the origin:

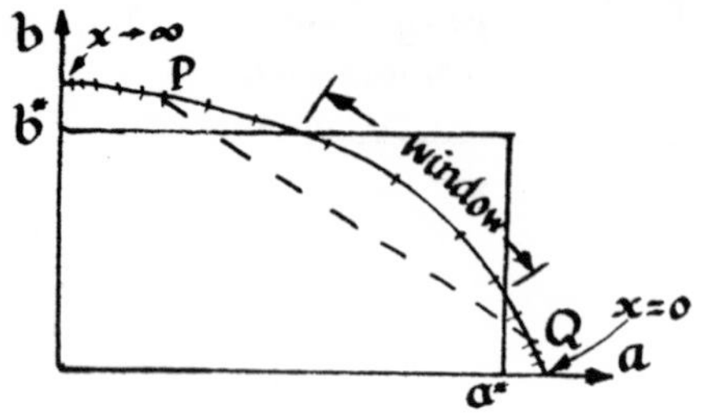

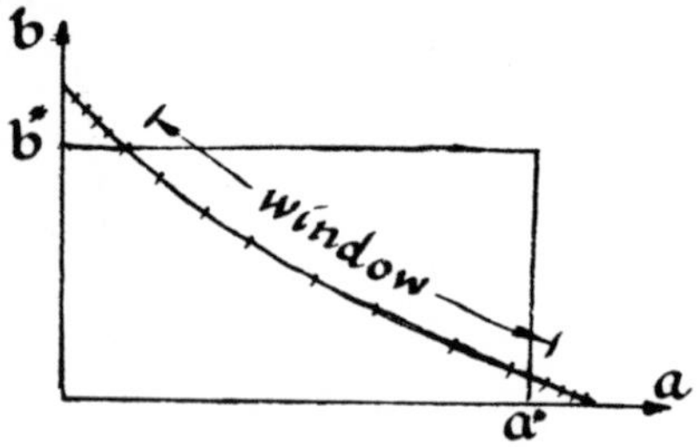

The window can again be read off, but now it becomes clear that it is easier to get away from both limitations when the curve is convex to the origin. On the other hand if the requirements can be met in the mean, they can more easily be met if the curve is concave to the origin for then the result of operating in slow oscillation between two such points as P and Q is represented a point on the chord PQ, which opens up a wider window for oscillatory operation.

5.2 Extensions of models.

In this section I want to raise a rather speculative question. When, if ever, is it profitable to extend consideration of a model into physically unreal regions? The answer could be, 'Never', since the physically correct regions should define itself and be closed. Thus, for example, with

$$V\frac{dc}{dt} = q(c_f-c)-VA_e{}^{-E/RT}c$$

$$VC_p\frac{dT}{dt} = qC_p(T_f-T)-h(T-T_{cf}) + (-\Delta H)VAe^{-E/RT}c$$

the region

$$0\le c\le c_f,\ 0\le T$$

is the physically appropriate region. Moreover we see that $(dc/dt)<0$ on $c = c_f$, but $(dc/dt)>0$ on $c = 0$; while $(dT/dt)>0$ for $T=0$ and (dT/dt) is negative for sufficiently large T.

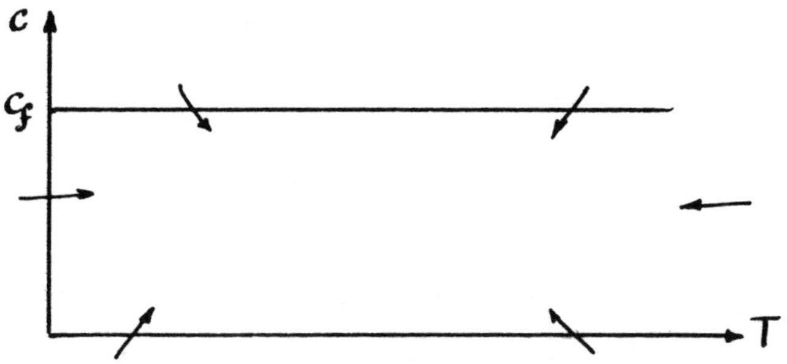

Thus the region of physical interest is closed and one could argue that attention should be confined to it. One cannot argue that attention can be confined to any sub-region on the grounds that only such a region is of practical significance unless it also is similarly closed. Thus D_1 would be an admissible confinement, but not D_2. But the equations themselves are indifferent toward physical reality, let alone practical significance,

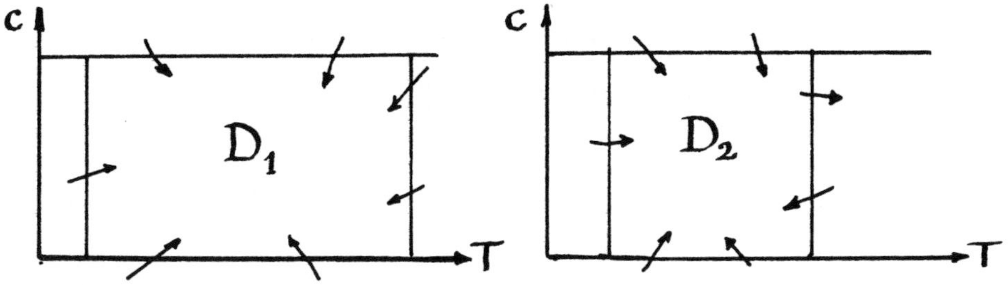

and obtain throughout the whole c,T plane. There is in this case the 'monstrous steep of Montmorency' at T=0 which we might expect to be a natural barrier.

But the enlargement of the variables beyond the bounds of physical reality is related to a similar extension of the parameters. In the above equations V, q, c_f, A, E, C_p, h, T_f and T_{cf} are all positive and only ΔH can change sign. In the non-dimensional form of the equations (cf. Sec. 4.3)

$$\frac{du}{d\tau} = 1 - u - \alpha u e^{-1/v} \tag{5.6}$$

$$\frac{1}{\delta}\frac{dv}{d\tau} = \bar{v} - v + \alpha\zeta u e^{-1/v} \tag{5.7}$$

the only parameter which can change sign is ζ.

However, there are at least two prima facie reasons why it might be desirable to extend the range of the parameters. First, the extension will allow us to see the system more comprehensively and understand its movements better. Second, it may suggest connections with other systems or even suggest new ones. Thus a stirred fermentor with growth by a logistical curve would have an equation

$$V\frac{dc}{dt} = -qc + kc(c_m - c)$$

with V, q, k and c_m positive. However allowing k to be negative gives the equation for a stirred tank with an autocatalytic reaction. Let us examine the steady state of the stirred tank in more detail.

The equations for u and v at steady state are

$$\begin{aligned} 1 - u - \alpha u e^{-1/v} &= 0, \\ \bar{v} - v + \alpha\zeta u e^{-1/v} &= 0; \end{aligned} \tag{5.8}$$

these can be reduced to the single equation which has been cited as (5.1) but is repeated here:

$$\frac{1}{\zeta}(v - \bar{v}) = \frac{\alpha e^{-1/v}}{1 + \alpha e^{-1/v}} = Z(v;\alpha) \tag{5.9}$$

The left hand side is a straight line, the right a curve with a discontinuity at $v = 0$ and depending on only one parameter. There is a discontinuity at $v = 0$, since $z(0-0;\alpha) = 1$ but $z(0+0;\alpha) = 0$. The general form of the function is given on p. 117 for positive α and we see immediately that an arbitrary straight line can have 0, 1, 2 or 3 intersections. The forms for

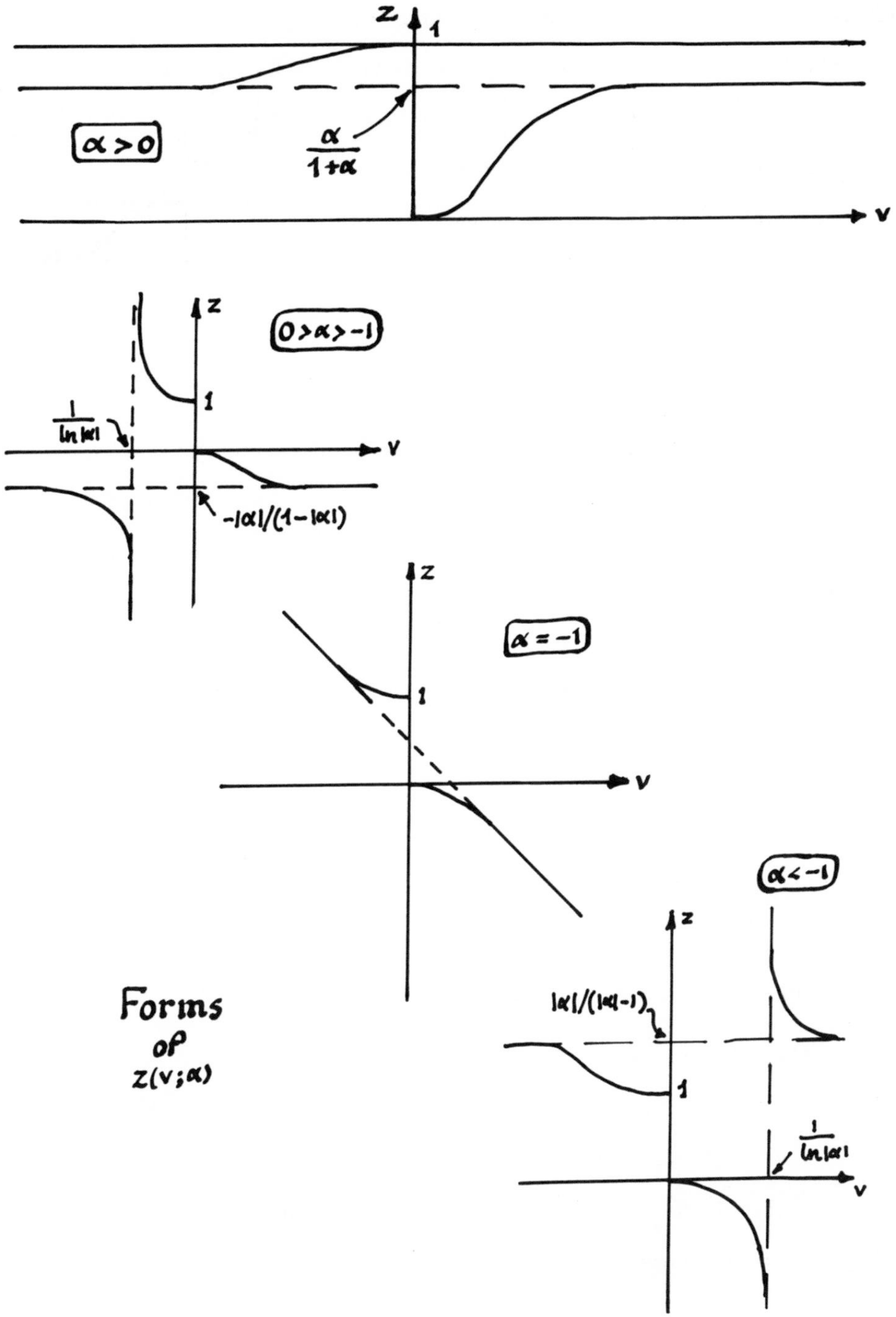
z
1
α > 0
α/(1+α)
v
z
0 > α > −1
1
1/ln|α|
v
−|α|/(1−|α|)
z
α = −1
1
v
α < −1
z
|α|/(|α|−1)
1
1/ln|α|
v
Forms of z(v;α)

negative α are also shown and up to three intersections are possible unless the branches of the curves are so disposed that an intersection of the form shown on the right is possible. We shall eliminate this possibility later.

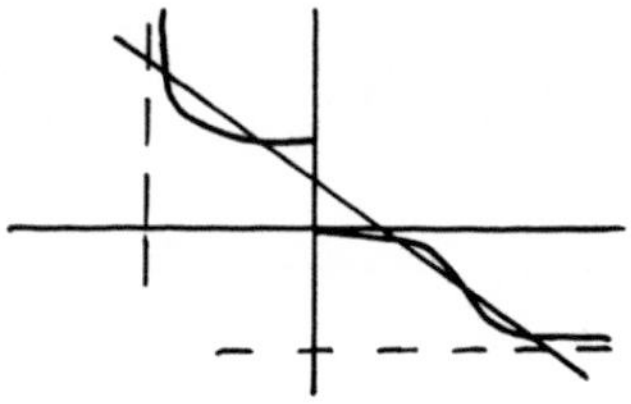

The function $z(v;\alpha)$ has the properties

$$z(v;\alpha) \to \alpha/(1+\alpha) \text{ as } v \to \infty$$

$$z'(v;\alpha) = z(1-z)/v^2 \qquad (5.10)$$

$$z''(v;\alpha) = z(1-z)(1-2z-2v)/v^4$$

so the inflections points are always located on the line

$$v + z = 1/2 \qquad (5.11)$$

which is the asymptote when $\alpha = -1$. We notice a certain rotational symmetry about $z = 1/2$, $v = 0$ and in fact

$$z(v;\alpha) + z(-v;\alpha^{-1}) = 1$$

or (5.12)

$$z(v;\alpha) - 1/2 = -\{z(-v;\alpha^{-1}) - 1/2\}$$

Thus if $v-\bar{v}=\zeta z(v;\alpha)$ has a solution $v = w$ then $v-\bar{v}' = \zeta' z(v;\alpha')$ will have the solution $v = -w$ provided

$$\alpha' = \alpha^{-1}, \ \zeta' = \zeta \ \text{ and } \ \bar{v}' = -\bar{v}-\zeta.$$

To show this we write $\alpha' = \alpha^{-1}$ and

$$v-\bar{v}' = \zeta' z(v;\alpha^{-1}) = \zeta' - \zeta' z(-v;\alpha)$$

or by replacing v by $-v$,

$$v + \bar{v}' + \zeta' = \zeta' z(v;\alpha).$$

Thus in principle we can find out everything by studying the cases for $|\alpha| \le 1$. It also suggests a transformation of parameters to

$$\alpha, \zeta \ \text{ and } \ w = \bar{v} + (\zeta/2) \qquad (5.13)$$

will introduce greater symmetry since the diagrams will be invariant under $\alpha \to \alpha^{-1}$, $\zeta \to \zeta$, $w \to -w$.

We will trace the pattern of roots of eqn. (5.9) in two cases: $0<\alpha<1$ and $0>\alpha>-1$. For $\alpha>0$ there are two critical points for $\bar{v}$, namely the intersection on the v-axis of the tangents at the points of inflection, say V_+ and V_-.

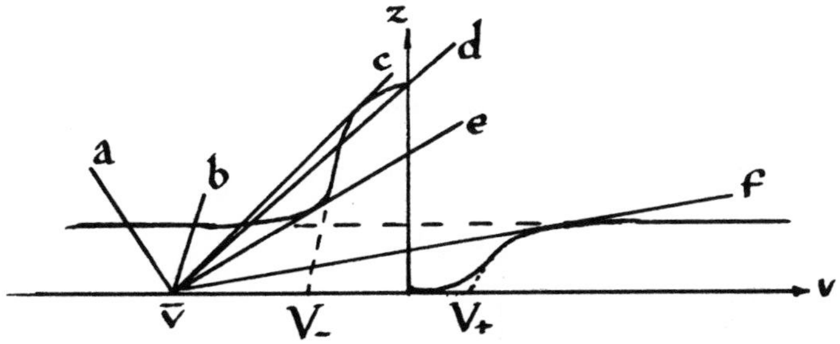

As ζ goes from $-\infty$ through 0 to ∞ the line through $\bar{v}$ swings clockwise from the horizontal through the vertical to the horizontal. Such a sequence for $\bar{v}<V_-$ is shown above as $\bar{v}a$, $\bar{v}b$,...$\bar{v}f$. Until the positive slope of $\bar{v}e$ is reached the intersection of lines such as $\bar{v}a$ and $\bar{v}b$ is unique. A line between $\bar{v}c$ and $\bar{v}d$ intersects thrice, but beyond $\bar{v}d$ only twice until $\bar{v}e$ is reached. $\bar{v}d$ is the line with $\zeta = -\bar{v}$ and, as $\bar{v}\to-\infty$, $\bar{v}c$ and $\bar{v}d$ tend to coincide. On the other hand the slope of $\bar{v}e$ is the value of z' at the point of tangency E and this tends to $\alpha/(1+\alpha)^2y^2$ as y, the abscissa of E, tends to $-\infty$. Thus for $y\to-\infty$

$$\zeta\to(1+\alpha)^2y^2/\alpha$$

and since $z\to\alpha/(1+\alpha)$, we also have

$$\zeta\to(y-\bar{v})(1+\alpha)/\alpha$$

and hence asymptotically we have the relation

$$\alpha\zeta = [\alpha\zeta + (1+\alpha)\bar{v}]^2.$$

When the point $\bar{v}$ is close to V_- there is only a slight difference between the slopes of $\bar{v}c$ and $\bar{v}e$. The region of two or three intersections corresponding to lines between $\bar{v}c$ and $\bar{v}e$ is traced out by computing the slope of the tangent and its intersection with the v-axis, i.e. parametically by letting v run from $-\infty$ to 0 in

$$\zeta = v^2/z(1-z), \quad \bar{v} = v - z/z' = v(1-v-z)/(1-z) \tag{5.14}$$

Similarly the region of two intersections for a line of lesser slope than $\bar{v}f$ is traced out by letting v run from 0 to ∞. If $\bar{v}$ is positive but less than V_+ there are three intersections. Thus the pattern of multiplicity in the $\zeta\,\bar{v}$ plane is as below

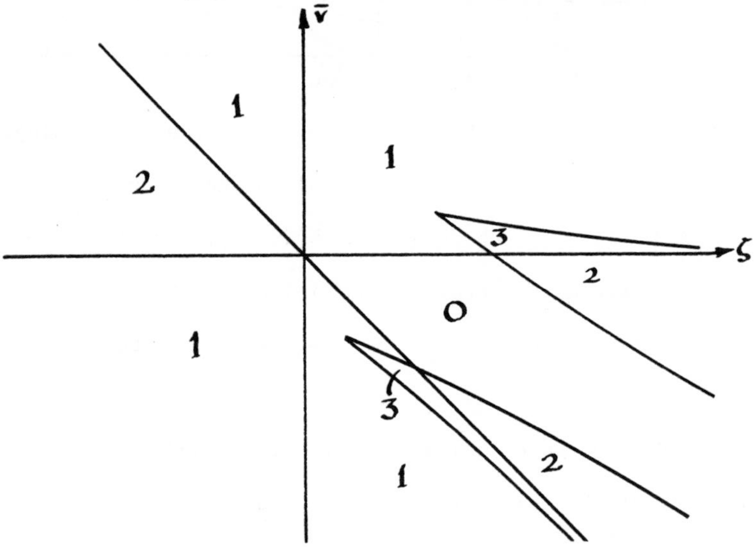

As α varies the cusps at the tip of the regions of triplicity move on the hyperbola $\bar{v}^{-2} + \bar{v}\zeta = \zeta/4$. There is a measure of symmetry about the line $\bar{v} = -\zeta$ and this confirms the previous impression that $w = \bar{v} + 1/2\zeta$ may have merit as a parameter. In the ζ, w-plane we have

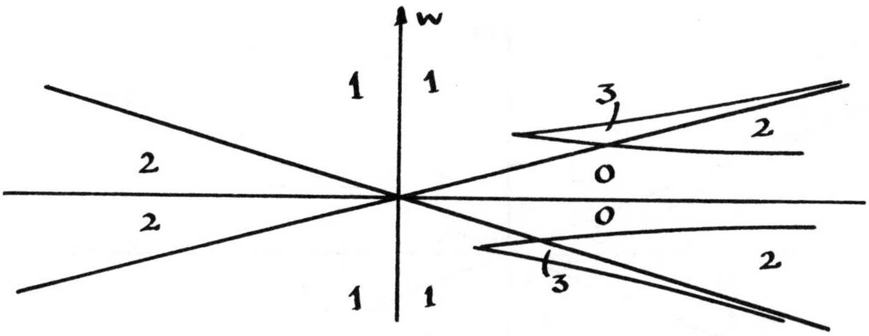

In these coordinates the locus of cusps is the hyperbola

$$4(\zeta+1/2)^2 - 16w^2 = 1.$$

The pattern for α^{-1} is obtained from that for α by rotation about the ζ axis.

For negative α we can use a similar argument first tracing out the ζ, $\bar{v}$-locus by eqn. (5.14). The correspondence is best shown separately.

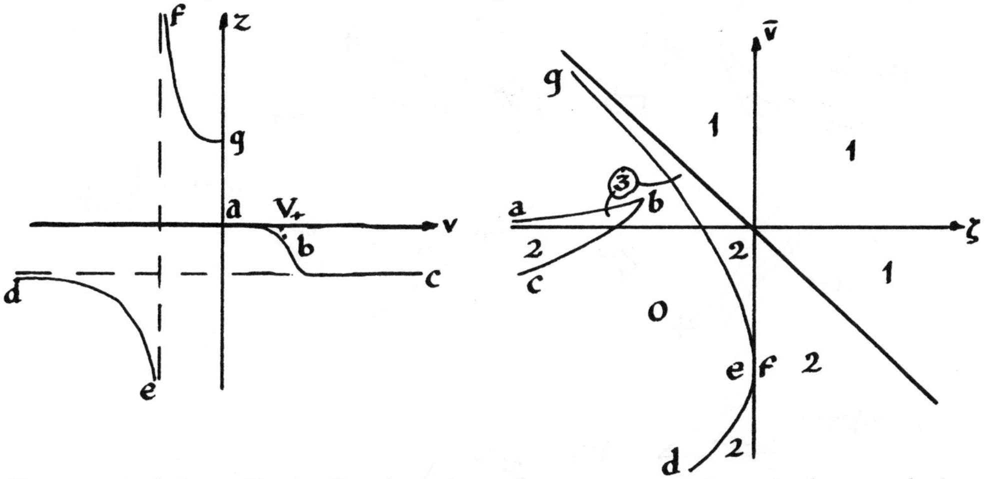

The symmetrizing effect of using ζ and w as parameters is less marked but is shown below and the form for α^{-1} is obtained by reflection in $w=0$.

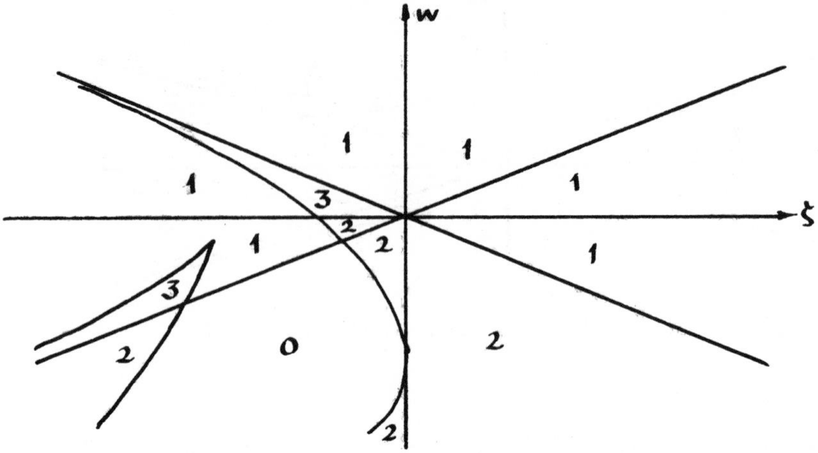

There are echoes here of the canonical catastrophes but there seems no obvious correspondence. Thus in the movement of the cusp in the case of negative α we have:

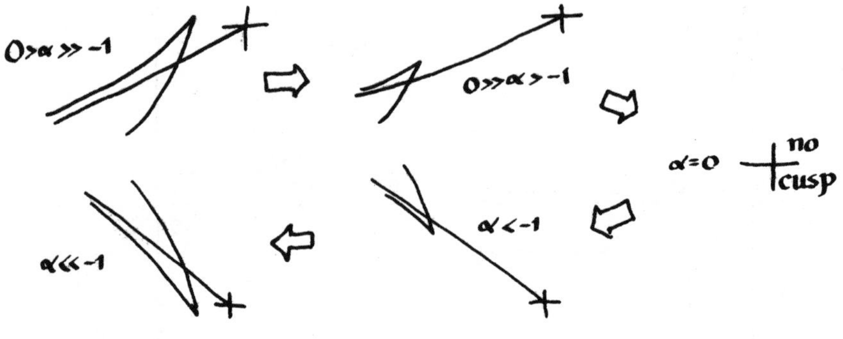

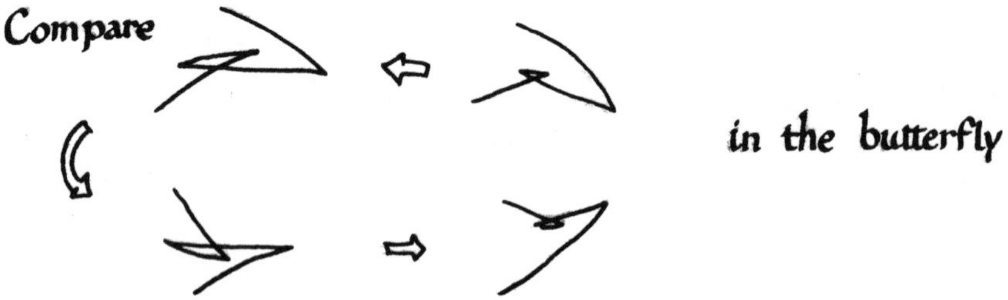

Apart from getting an enlarged view of the system of equations there may be links with similar equations in other areas. In this case, there is an analogue from statistical mechanics where Fermi-Dirac and Bose-Einstein statistics show complementary changes of sign. The mean number of particles in a state s is $\bar{n}_s, \varepsilon_s$ is its energy. Then Fermi-Dirac statistics give

$$\bar{n}_s = \left\{e^{\alpha+\beta\varepsilon_s}+1\right\}^{-1} = z\left(\frac{1}{\beta\varepsilon_s}\ ;\ e^{-\alpha}\right),$$

while Bose-Einstein give

$$\bar{n}_s = \left\{e^{\alpha+\beta\varepsilon_s}-1\right\} = z\left(\frac{1}{\beta\varepsilon_s}\ ;\ -e^{-\alpha}\right).$$

Since β is the chemists' abbreviation for $1/kT$, $1/\beta\varepsilon_s$ is a dimensionless temperature v.

5.3 Observable quantities.

Returning from the speculative to the practical, a word should be inserted about the value of using observable quantities. What is meant by this is best illustrated by an example. In Sec. 4.4 we looked at the equations for diffusion and reaction in a catalyst pellet

$$\nabla^2\xi + R(\xi) = 0 \quad \text{in} \quad \Omega,$$
$$\xi = 0 \quad \text{on} \quad \partial\Omega. \tag{4.66 bis}$$

To see the application of observable quantities we need a slightly different form of the equations. Now ξ has the curious dimensions of moles per unit length per unit time and it needs to be made dimensionless by being divided by the product of a characteristic value of a diffusion coefficient and concentration. Let these be D^* and c^* and set

$$w = \xi/D^*c^* \tag{5.15}$$

Also let us write $R(\xi)$ in the form

$$R(\xi) = r[c_{1s} + (\alpha_1/D_1)\xi, \ldots T_s(-\Delta H)/k_e)\xi]$$

$$= r[c_{1s} + (\alpha_1 D^*/D_1)c^*w, \ldots, T_s + \{(-\Delta H)D^*c^*/k_e\}w] \quad (5.16)$$

$$= r(c_{1s}, \ldots T_s)P(w)$$

so that P is the ratio of the reaction rate to its value under surface conditions. Finally if the variables in the Laplacian are made dimensionless by dividing by a characteristic length L^* we have

$$\nabla^2 w + \phi^2 P(w) = 0 \text{ in } \Omega \quad (5.17)$$

$$w = 0 \quad \text{on } \partial\Omega \quad (5.18)$$

where

$$\phi^2 = L^{*2}\frac{r(c_{1s}, \ldots T_s)}{D^*c^*} \quad (5.19)$$

This last parameter (often known as the Thiele modulus in honour of one of the pioneers in this field) is a measure of the ratio of the reaction rate to the diffusion rate. There will of course be other parameters in $P(w)$ but, supposing these to be constant, the solution of eqn. (5.17) will be a function of ϕ.

Now there is a very important functional of the solution, namely

$$\eta = \frac{1}{V_\Omega}\iiint_\Omega P\ (w)dV. \quad (5.20)$$

the average value of the reaction rate in the pellet as a ratio of the surface reaction rate. This so-called "effectiveness factor" is a function of the parameter ϕ and wraps up the practical implications of the solution very neatly. As $\phi \to 0$, $\eta \to 1$, for eqn. (5.17) becomes Laplace's equation and, with the boundary condition (5.18), the solution is $w=0$; but $P(0)=1$ so $\eta=1$. Physically this makes sense, for $\phi \to 0$ means that diffusion is rapid in comparison with reaction and so the surface conditions prevail everywhere. As $\phi \to \infty$, $\eta \to 0$ since, with increasingly rapid reaction and slower diffusion, the interior of the pellet is starved of reactants and ineffective. In fact a singular perturbation analysis shows that η is inversely proportional to ϕ for large ϕ. The log-log plot of $\eta(\phi)$ has the form shown in the simplest

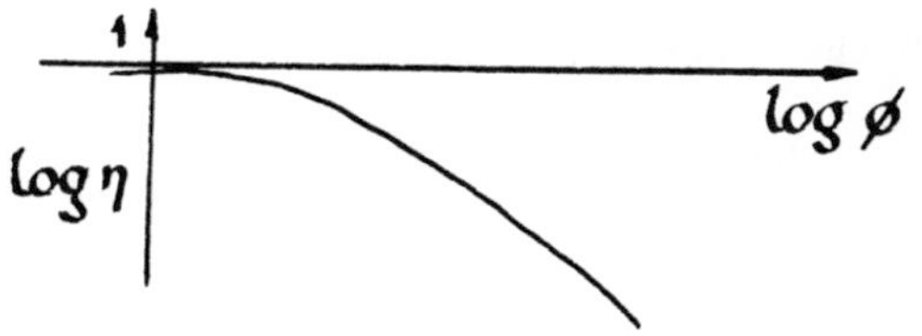

cases. For example if $P(w)=1-w$, $\eta(\phi)=(\tanh\phi)/\phi$ when Ω is a flat plate of thickness $2L^*$ with sealed edges; it is $\eta(\phi) = (3/\phi^2)(\phi\coth\phi-1)$ for a sphere of radius L^*.

Now it is all very well to know $\eta=\eta(\phi)$ but when the reaction rate is measured it is not $r(c_{1s},\ldots T_s) = r_s$ that is observed, but rather $\eta r_s = r_{obs}$ since the intrusive effects of diffusion are present in the measuring process. It is therefore useful to define an observable parameter

$$\Phi^2 = L^{*2} r_{obs}/D^*c^* = \eta\phi^2 \tag{5.21}$$

and to plot η as a function of Φ rather than of ϕ. We notice that Φ and ϕ are virtually the same when they are small but that Φ is proportional to the square root of ϕ, when they are large. Thus the graph of $\eta(\Phi)$ is a little steeper and so defines a little better than $\eta(\phi)$ a value

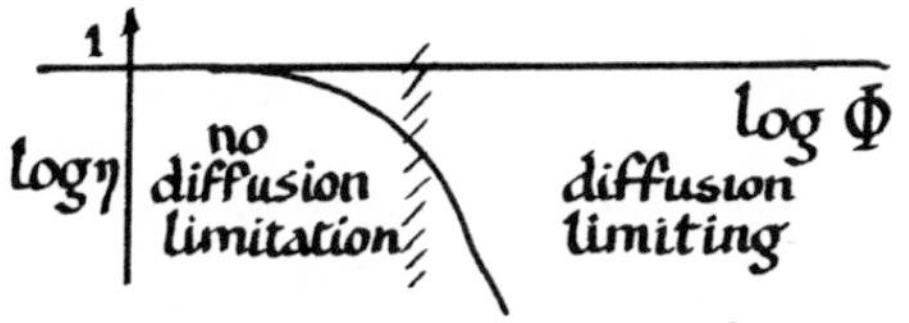

of Φ above which diffusion limitation can be regarded as serious. There is of course something arbitrary about such a critical value (it might for instance, be the value at which $\eta = 0.8$, or $\eta = 0.9$ if one wanted to be more cautious), but it can be a very sensible arbitrariness and once determined can be correlated with other parameters. Bischoff [30] has shown how to normalize ϕ so that a sensible critical value is $\phi=1$.

The notion of a normalized modulus is an elegant one and often worth cultivating. For example, we have noted that the asymptotic form of η is inversely proportional to ϕ. The constant of proportionality depends on the shape of Ω and on the kinetics. However if w is scaled so that $P(1)=0$ and L^* is chosen so that

$$L^* + \frac{V}{S}\left\{\int_0^1 P(w)\,dw\right\}^{1/2}, \tag{5.22}$$

where V is the volume and S the surface area of Ω, then the constant of proportionality is 1 (i.e. $\eta \sim \phi^{-1}$) for all shapes and kinetics. The basis for this statement lies in the singular perturbation solution of eqn. 5.17 and will not be pursued here, but it can readily be appreciated that when ϕ is large only a thin layer beneath the surface is contributing to the reaction and hence the surface area if important. Details can be found in [14].

5.4 Comparison of models and prototypes and of models among themselves.

When a model is being used as a simulation an obvious comparison can be made between its predictions and the results of experiment. We are favourably impressed with the model if the agreement is good and if it has not been purchased at the price of too many empirical constants adjusted to fit the data. If the parameters are determined independently and fed into the final model as fixed constants not to be further adjusted, then we can have a fair degree of confidence in the data and in the model. Both model and data have their own integrity, the former in the relevance and clarity of its hypotheses and the rigour and appropriateness of its development, the latter in the carefulness of the experimenter and the accuracy of the results. But these virtues do not only inhere in the possessors they also gain validity from the other. Thus, as Truesdell remarks, in applied mathematics rigour is of the essence, for the comparison can have no meaning if the model has not been handled properly. Similarly data must be of a certain degree of accuracy or it has no ability to prove (i.e. test) a theoretical viewpoint. Thus the attitude of never believing an experiment until its confirmed by theory has as much to be said for it as that which never believes a theory before its confirmation by experiment.

In the comparison of theory with experiment an array of statistical tools is available and should be used. Thus not only can the fitted constants be chosen in some best sense (e.g. least squares) but it is not difficult to find also the covariance matrix of the estimates and hence detect any hidden sensitivities. One danger that is easy to overlook is the existence of hidden constancies that will give spurious values. Thus the temperature rise in an adiabatic bed is a measure of the reaction rate which will be a function of the mean temperature. But if the inlet temperature is virtually constant, the observed pairs of temperature rise and mean temperature will be perfectly correlated by a straight line whatever the functional relationship between them. This straight line says no more than that the mean temperature equals the inlet plus half the rise. The classic correlation between the intelligence of the children and the drunkenness of the parents which so confounded temperance societies years ago--until it was discovered that all the data came from schools in the east end of London--is another illustration of a data base too narrow to test a model.

In discriminating between models it is not entirely satisfactory to fit the constants of each and choose the better fitting model. For one thing there may be little to choose between the goodness of fit in the two cases. We are on much firmer ground if the two models can be presented in such a way that they have qualitatively different behaviour. Tanner [171] has tried plotting data on the intermediates of a complex reaction in such a way that they fall on a loop which in some models is traversed clockwise and in others anti-clockwise.

A field in which qualitative behaviour of models has been used discriminatingly is the study of oscillations in chemical reaction. If a certain reaction is known to give oscillations under certain conditions then any mechanism that is incapable of giving oscillations under these conditions is ruled out. Sheintuch and Schmitz have reviewed this subject very thoroughly and examined the models for the oxidation of carbon monoxide in the light of this criterion, [163]; see also Eigenberger [61] and Luss and Pikios [124]. They find that a more than usually detailed account of the mechanism is needed. Unless the lack of uniformity of the catalyst surface, the affects of the chemisorbed species, their variations of reactivity and the dependence of the activation energy on coverage are brought into consideration no oscillations can occur. This rules out seven of the thirteen cases they consider and some of the remaining possibilities are seen to be unlikely by the magnitude that various terms would have to have if oscillations were to occur. This is remarkable in that twelve of the thirteen cases can be adjusted to match the known form of the reaction rate expression for carbon monoxide oxidation.

This work of Schmitz and Sheintuch shows the power of mathematical modelling when combined with physical understanding, for not only are many possibilities eliminated, but the features that call for further investigation in the remaining candidate models are clearly brought out. The mention of it is a pleasant note on which to conclude this exploration of the craft of mathematical modelling.

Appendix A Longitudinal diffusion in a packed bed

Description of the system P.

A fluid flows through the interstices of a long cylinder which is packed with particles. Because of the variations in local velocity as the fluid passes around the particles, the eddying and wall effects, some molecules will pass through more quickly than others. In addition there is the molecular diffusion of the tracer molecules in the flow field. Hence, if the fluid is marked with a tracer which enters the bed at time $t = 0$ in a sharp pulse and the concentration of the tracer is measured as the stream leaves the bed,

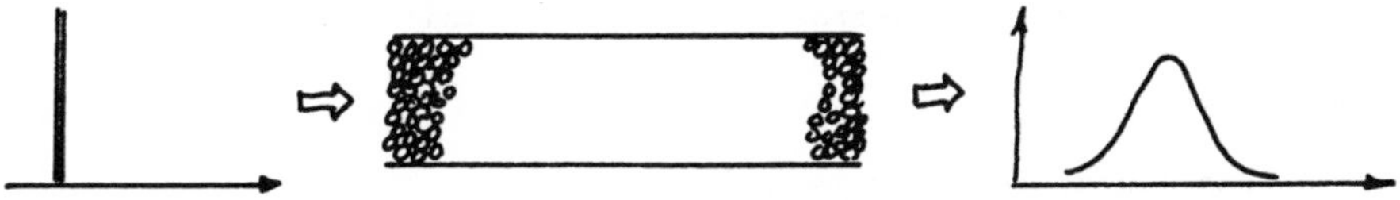

it will be found to have spread out into a diffuse band. If the input were perfectly instantaneous and the output $C(t)$ then

$$\mu = \int_0^\infty tC(t)dt / \int_0^\infty C(t)dt$$

would be the mean residence time of tracer molecules and

$$\sigma^2 = \int_0^\infty (t-\mu)^2 C(t)dt / \int_0^\infty C(t)dt$$

the variance of residence times. Clearly these two statistics of $C(t)$ give some idea of the dispersive effect of the interaction of these physical processes though the equations should be capable of yielding $C(t)$ itself. We denote by Ω the intersticial space through which the fluid flows, by $\partial\Omega$ the bounding surface of this region excluding $\partial\Omega_i$ and $\partial\Omega_o$, the inlet and outlet ends of the cylinder. Let the area of cross-section of the cylinder (and thus of $\partial\Omega_i$ and $\partial\Omega_o$) be A, q the volumetric flow rate, L the length and ε the fractional free space. The mean linear velocity of flow is thus $q/A\varepsilon = L/\Theta$, where Θ is defined by this equation as a residence time. The concentration of tracer at any point $\underset{\sim}{r}$ of the free space and time t is $c(\underset{\sim}{r},t)$; $C(t) = \iint_{\partial\Omega_o} c(r,t)\,dS$.

Hypotheses.

Physical laws and general conservation principles will be regarded as an underlying hypothesis, H_o. Then the specific hypotheses we need are as follows:

H_1: the fluid is incompressible and the flow steady.

H_2: its motion is governed by the Navier-Stokes equations.

H_3: the diffusion of the tracer obeys Fick's law.

H_4: the tracer does not penetrate the cylinder walls or enter the particles.

H_5: the mean linear velocity is uniform.

H_6: the dispersive effect is given by an effective longitudinal diffusion coefficient.

H_7: the movements of a tracer particle can be thought of in discrete time intervals during which it either moves forward by an increment of length or, being caught in an eddy, moves not at all.

Model Π_1.

This model is just the embodiment of the hypotheses H_o-H_4 using well-known equations; we shall not go into the derivation of the Navier-Stokes and similar equations but use them as needed; they are of course themselves derived by the principles outlined in Ch. 3. Thus, if $\underset{\sim}{v}(\underset{\sim}{r},t)$ is the velocity at some point $\underset{\sim}{r}$ in Ω, H_1 implies that it is really $\underset{\sim}{v}(\underset{\sim}{r})$ and the incompressibility that

$$\nabla \cdot \underset{\sim}{v} = 0 \tag{A1}$$

The Navier-Stokes equations are

$$(\underset{\sim}{v} \cdot \nabla)\underset{\sim}{v} = -\nabla\bar{p}/\rho + \nu\nabla^2\underset{\sim}{v}. \tag{A2}$$

and these are also written for a steady flow with $\bar{p}$ the pressure, ρ the density and ν the kinematic viscosity. The equation governing the dispersion of the tracer is not a steady equation but, with D the molecular diffusion coefficient, is

$$\frac{\partial c}{\partial t} + \underset{\sim}{v} \cdot \nabla c = D\nabla^2 c. \tag{A3}$$

These three equations are subject to the boundary conditions

$$\underset{\sim}{v} = 0 \text{ on } \partial\Omega,\ \bar{p} = \bar{p}_i,\ \underset{\sim}{v} = \underset{\sim}{v}_i \text{ on } \partial\Omega_i \tag{A4}$$

and if $\underset{\sim}{n}$ denotes the outward normal to the boundary of Ω,

$$\begin{aligned} &\underset{\sim}{n} \cdot \nabla c = 0 \text{ on } \partial\Omega \text{ or } \partial\Omega_o, \\ &\underset{\sim}{v}c - D\nabla c = \underset{\sim}{v}c_i \text{ on } \partial\Omega_i. \end{aligned} \tag{A5}$$

If the impulse of tracer put to the bed is perfectly sharp we might put

$$c_i = \delta(t)/q \tag{A6}$$

since then $\int_0^\infty\iint_{\partial\Omega} c_i\underset{\sim}{v}\cdot\underset{\sim}{n}\, dS\, dt = 1$. Initially the bed is free of tracer so

$$c(\underset{\sim}{r},o) = 0 \tag{A7}$$

<u>Model Π_2.</u>

The first model is so complicated as to be almost impossible of solution and clearly the complexity of the geometry is part of the problem. Suppose we take the drastic step of ignoring this complexity and say that the flow averages out to a virtually uniform velocity $U = q/A\varepsilon$ and that the dispersion is the sort of thing we would see if Fickian diffusion were imposed on this. The prototype has thus been modified by the hypotheses, H_5, H_6, to an equivalent continuum. The equations for this are for a concentration $c(x,t)$ which is a function of x the distance from the inlet and time. Then

$$\frac{\partial c}{\partial t} + U\frac{\partial c}{\partial x} = D_e \frac{\partial^2 c}{\partial x^2}, \quad 0 \leq x \leq L, \tag{A8}$$

where D_e is the effective diffusion coefficient. As boundary conditions we have

$$Uc - D_e \frac{\partial c}{\partial x} = Uc_i, \quad x = 0 \tag{A9}$$

and

$$\frac{\partial c}{\partial x} = 0, \quad x = L. \tag{A10}$$

Initially,

$$c(x,o) = 0. \tag{A11}$$

The measured output is

$$C(t) = c(L,t). \tag{A12}$$

Model Π_3.

In formulating Π_3 we take the opposite view of the physical system from the continuous analogue of P_2 and emphasize the discreteness of the packed bed. In particular the flow is a seaweed flow, squeezing between particles and bulging into the cavities. For example, in a rhombohedral blocked passage arrangement the overall fractional free volume is 26% but the free area perpendicular to the flow varies widely. Through a plane of centres of the spheres it is only 9% while if the region between one plane of centres and the next is divided into thirds the average free area is 41% in the middle third but only 18% in the other two. It is not entirely unreasonable therefore to regard each layer of particles as a cell in whose interstices, a thorough mixing takes place. The system P is thus replaced by P_3, a sequence of N cells each of volume V through which the stream passes. Equating the total free volume gives $NV = AL\varepsilon$. We will not be dogmatic about N at this point, though, if the analogy holds up we might expect N to be of the order of L/d_p where d_p is the particle diameter.

If $c_n(t)$ is the concentration of tracer in cell n

$$V \frac{dc_n}{dt} = q(c_{n-1}-c_n) \tag{A13}$$

with

$$V \frac{dc_1}{dt} = q(c_i-c_1) \tag{A14}$$

and

$$c_n(0) = 0 \tag{A15}$$

The observed quantity

$$C(t) = c_n(t). \tag{A16}$$

This model has been elaborated to consider radial as well as longitudinal dispersion in packed beds by Deans and Lapidus [53].

The connection between Π_2 and Π_3.

As is remarked in Ch. 1, there is no immediate connection between Π_2 and Π_3 for the discretization of Π_2 would not produce Π_3. They are connected only in the sense that they give comparable solutions. Thus if $c_i(t)$ is given the solution of Π_2 is

$$C(t) = \int_0^t c_i(t')p_2(L,t-t')dt' \tag{A17}$$

where

$$p_2(x,t) = U(\pi D_e t)^{-1/2}\exp - \frac{(x-Ut)^2}{4D_e t} - \frac{U^2}{2D_e}\exp\frac{Ux}{D_e}\operatorname{erfc}\frac{x+Ut}{2(D_e t)^{1/2}} \quad \text{(A18)}$$

In particular if $c_i(t)$ is the delta function input, the output is $p_2(L,t)$ and $\int_0^t p_2 dt=1$. The distribution of residence times is almost Gaussian and the mean and variance of residence times are

$$\mu_2 = \frac{L}{U}\left[1 + \frac{1}{P}\right] \quad \text{(A19)}$$

$$\sigma_2^2 = \frac{2DL}{U^3}\left[1 + \frac{3}{4P}\right] = \frac{L^2}{U^2 P}\left[1 + \frac{3}{4P}\right] \quad \text{(A20)}$$

where $P = UL/2D_e$.

The solution for Π_3 is

$$C(t) = \int_0^t c_i(t')p_3(t-t')dt' \quad \text{(A21)}$$

where

$$p_3(t) = \frac{1}{(N-1)!}\left[\frac{q}{v}\right]^N t^{N-1}e^{-qt/v} \quad \text{(A22)}$$

This is the Poisson distribution which is also asymptotically Gaussian and whose mean and variance are

$$\mu_3 = NV/q \quad \text{(A23)}$$

$$\sigma_3^2 = NV^2/q^2 \quad \text{(A24)}$$

Since the total volume of the cells, NV, should obviously equal the total free space, $LA\varepsilon$, and $q = A\varepsilon U$, we see that

$$\mu_3 = NV/q = L/U = \Theta$$

which is approximately equal to μ_2 if $P>>1$. But if we equate μ^2/σ^2 in the two cases we see that

$$N = P(1 + \tfrac{1}{P})^2/(1 + \tfrac{3}{4P})$$

$$\doteqdot P + \tfrac{5}{4}\,.$$

As we have seen, there are physical grounds for thinking that N should be of the order of L/d_p which is quite large; hence $P \doteq N >> 1$ and the means and variances of the two models are approximately the same. If $P = N = L/d_p$ then the so-called particle Peclet number $Pe = Ud_p/D_e = 2$ which is in good agreement with experiment. An alternative way of establishing the connection between Π_2 and Π_3 by the common Gaussian approximation to p_2 and p_3 was discussed by Amundson and Aris [5].

Model Π_4.

A rather different model is obtained by taking a disjointed view of the tracer movement and saying that sometimes it moves forward with the stream and at others it is caught in an eddy and stays in virtually the same place. This is the crudest of random walk assumptions, embodied in H_7, and obviously could be elaborated by giving a distribution of lengths over which the movement might take place. Suppose that in each interval of time τ the particle either moves forward a distance δ or remains where it is. The probability of the first event is p and of the second is $q = 1-p$. During a time $t = M\tau$ there have been M such "choices" and, if $L = N\delta$, the particle will emerge if N of these have been to move forward. The probability of this is

$$p_4 = \binom{M}{N} p^N q^{M-N} \tag{A25}$$

Thus

$$C(t) = \binom{t/\tau}{L/\delta} p^{(L/\delta)} q^{(t/\tau)-(L/\delta)} \tag{A26}$$

It is well known that if

$$z = \frac{N-Mp}{(Mpq)^{1/2}} = \frac{L-(\delta/\tau)tp}{\delta(tpq/\tau)^{1/2}} \tag{A27}$$

then

$$C(\tau) \sim \frac{1}{(2\pi tpq/\tau)^{1/2}} \exp - \frac{(L-(\delta p/\tau)t)^2}{2(\delta^2 pq/\tau)t} \tag{A28}$$

as N and $M \to \infty$ and $z^3 M^{-1/2} \to 0$. Comparing this with (A18) we see that δ, τ and p should satisfy

$$\delta p/\tau = U, \ \delta^2 pq/\tau = D_e. \tag{A29}$$

Thus if δ is of the order of d_p and $Pe = Ud_p/D_e = 2$ we have $q = p = 1/2$, which is reasonable enough.

<u>Notation for the models Π.</u>

A	cross-sectional area of bed
$C(t)$	outcoming average concentration
$c(\underset{\sim}{r},t)$	concentration at position $\underset{\sim}{r}$ and time t
$c_i(t)$	input concentration
$c_n(t)$	concentration in n^{th} cell
D	molecular diffusion coefficient of tracer
D_e	effective or equivalent diffusion coefficient in Π_2
d_p	particle diameter
L	length of packed bed
M	number of time increments in Π_4
N	number of cells in Π_3 or space increments in Π_4
P	$UL/2D_e$
Pe	Ud_p/D_e
p	probability of movement in Π_4, $q = 1-p$
p_2, p_3	residence time probability densities in Π_2 and Π_3
$\bar{p}$, $\bar{p}_i$	pressure, inlet pressure in Π_1
q	volumetric flow rate
$\underset{\sim}{r}$	reactor of position in free space of packed bed

t	time
U	linear velocity
V	volume of cell
$\underset{\sim}{v}$	reactor of velocity in interstices of bed
x	distance from inlet
δ	length of step in Π_4
ε	fractional free volume of bed
Θ	residence time
μ_i	mean residence time in Π_i, $i = 2,3$
ν	kinematic viscosity
ρ	density
σ_i^2	variance of residence times in Π_i, $i = 2,3$
τ	time increment in Π_4
Ω	free space of packed bed
$\partial\Omega$	boundary of Ω except for $\partial\Omega_i$ and $\partial\Omega_o$
$\partial\Omega_i$, $\partial\Omega_o$	inlet and outlet boundaries

Appendix B The coated tube chromatograph and Taylor diffusion

Description of the systems C and D.

The inside of a long, cylindrical tube is coated with a thin retentive layer. A carrier gas flows through the tube and molecules of a tracer solute are convected by it, they diffuse and, if they reach the wall, may pass into the retentive layer and spend some time there. Different tracer solutes with different affinities for the retentive layer will spend different proportions of their time in this stationary phase and so peaks of different solutes are separated emerging at different times. (The analogy is sometimes made with a stream of soldiers on a long road lined with pubs: the teetotalers will arrive at the end of the road first, followed by the temperate and the topers--the dipsomaniacs may never make it.) But the diffusion, the variations of velocity across the tube and the rate of partition between the phases will all contribute to the spreading out of an initially sharp peak. Our interest is to account for the mean speed of the peak and to understand how each factor affects its spread.

There is also the special subordinate case when there is no retentive layer (system D). Here attention is focussed entirely on the interaction of diffusion and convection. This is the so-called Taylor diffusion problem first successfully analyzed by Sir Geoffrey Taylor [174]. Lateral diffusion prevents the solute from travelling with any one streamline and counteracts the spreading effect of the wide variation of flow rates. Thus there is a Taylor diffusion coefficient which is inversely proportional to the molecular diffusion coefficient. An immense literature on this problem now exists which is it not our purpose to summarize here: an early summary was given by Taylor [175] and a later one by Gill and Nunge [75] but there has not been any survey of the most recent developments. Extensions to the chemical reactor where reaction takes place in the fluid, in the retentive layer or in both have also been proposed.

Hypotheses of systems C and D.

Again H_o will be taken to embrace the underlying scientific laws and the following are the specific assumptions that are introduced to define various models.

H_1: the tube in infinite in both directions

H_2: the tube stretches from the origin, $x=0$, to infinity

H_3: the tube is finite, $0 \leq x \leq L$.

H_4: the velocities in the cylinder, $0=r_o \leq r \leq r_1$, and the annulus, $r_1 \leq r \leq r_2$, are functions only of r, the distance from the axis say $U_i \phi_i(r), i=1,2$, where U_i is the mean velocity in the region.

H_5: the annular region (retentive coating) is stationary.

H_6: the coating is thin, $(r_2-r_1) << r_1$.

H_7: there is no coating.

H_8: the rate of exchange between the two regions is proportional to the difference $(c_2-\alpha c_1)$, where c_1 and c_2 are the concentrations in the cylinder and annulus and α a constant.

H_9: the diffusion coefficients D_i, $i = 1,2$, are constant (a more general hypothesis is made in Aris [11] but we need not reach for too great a generality).

H_{10}: there is no flow across the axis, $r = r_o=0$, or the outer wall, $r=r_2$.

H_{11}: sufficient insight into the problem is to be gained from the temporal evolution of the moments of the distribution in space.

H_{12}: sufficient insight is to be had from calculating the mean concentration across the tube and determining an effective speed and dispersion coefficient.

H_{13}: longitudinal diffusion is unimportant.

H_{14}: the flow profile is parabolic $\phi_1(r) = 2[1-(r/r_1)^2]$.

H_{15}: diffusion plays no role.

<u>The most general model Γ_1.</u>

Clearly some of the foregoing hypotheses are mutually contradictory and will be used to define different case and sets of boundary conditions. The basic equations of convection and diffusion (i.e. H_0) with H_4 and H_9 give

$$\frac{\partial c_i}{\partial t} + U_i\phi_i(r)\,\frac{\partial c_i}{\partial x} = D_i\left\{\frac{\partial^2 c_i}{\partial x^2} + \frac{1}{r}\frac{\partial}{\partial r}\left[r\,\frac{\partial c_i}{\partial r}\right]\right\}, \quad i = 1,2 \tag{B1}$$

The hypothesis H_{10} gives us the boundary conditions

$$D_1\,\frac{\partial c_1}{\partial r} = 0, \quad r = 0, \tag{B2}$$

and

$$D_2\,\frac{\partial c_2}{\partial r} = 0, \quad r = r_2, \tag{B3}$$

while H_8 gives

$$D_1\,\frac{\partial c_1}{\partial r} = D_2\,\frac{\partial c_2}{\partial r} = k(c_2-\alpha c_1), \quad r = r_1. \tag{B4}$$

The initial conditions have to be specified and are

$$c_i(x,r,o) = \chi_i(x,r), \quad i = 1,2. \tag{B5}$$

We have not said anything yet as to the boundary conditions with respect to x. If H_1 is asserted we need only add

$$c_i(x,r,t) \text{ is finite as } x \to \pm\infty \tag{B6}$$

and the problem is complete. Γ_1 is the set of equations (B1) - (B6).

It proves advantageous in the analysis of this problem to have a moving origin, so that alternative equations to (B1) are

$$y = x-Vt \tag{B7}$$

$$\frac{\partial c_i}{\partial t} + (U_i\phi_i(r)-V)\,\frac{\partial c_i}{\partial y} = D_i\left\{\frac{\partial^2 c_i}{\partial y^2} + \frac{1}{r}\frac{\partial}{\partial r}\left(r\,\frac{\partial c_i}{\partial t}\right)\right\}. \tag{B8}$$

This is not a new model but a preliminary modification, Γ_1'.

<u>The models Γ_2 and Γ_3.</u>

The doubly infinite tube is not a very accurate model of the practical situation. It can be realized by a long capillary tube with the variation of tracer concentration confined to a comparatively narrow range far from the ends. Indeed experiments have been done by flashing light on a short section of a tube through which the flow of a light sensitive fluid passes. Such a flash at $t=0$ established the initial concentration distribution χ and the conditions of Γ_1 apply very precisely. A far more realistic situation however is afforded by replacing H_1 by H_2. It then becomes necessary to replace the condition of finiteness as $x \to -\infty$ by an inlet condition at $x = 0$. If U_1 and U_2 are both positive and if it is felt that a solute fed to the plane $x = 0$ (for example from a reservoir) must enter the system then

$$D_i\,\frac{\partial c_i}{\partial x} + U_i\phi_i(r)\,\{c_i-c_{if}\} = 0,\ i=1,2 \tag{B9}$$

at $x = 0$, where c_{if} is the feed concentration.

If $U_2 < 0$, as when a thin film runs down the wall of a vertical cylinder countercurrent to an upward flow of gas then we have to invoke H_3 since the countercurrent film must be specified at its inlet and write

$$D_i \frac{\partial c_1}{\partial x} + U_1\phi_1(r)\ \{c_1 - c_{1f}\} = 0,\ x=0,\ U_1>0,$$
$$D_2 \frac{\partial c_2}{\partial x} + U_2\phi_2(r)\ \{c_2 - c_{2f}\} = 0,\ x=L,\ U_2<0 \qquad \text{(B10)}$$

With the semifinite tube there is still the requirement of finiteness as $x\to\infty$, but for the finite tube another condition is required. These are often written

$$\frac{\partial c_1}{\partial x} = 0,\ x=L;\ \frac{\partial c_2}{\partial x} = 0,\ x=0. \qquad \text{(B11)}$$

<u>The moment model Γ_4.</u>

The important characteristic of all dispersion situations is that there is a certain movement of the centre of gravity of the solute distribution and a steady increase in its spread. It follows that some insight may be gained by studying the temporal evolution of the spatial moments. If we had knowledge of <u>all</u> the moments then, with suitable restrictions, we should have knowledge of the whole distribution (see e.g. [162]), but, apart from the unlikely event of there being a special form of solution, this demands solution of an infinite number of equations and so is a worse task than the solution of Γ_1 itself. Of course, our hope is that the first two or three moments will tell us all we want to know.

Let

$$c_i^{(p)}(r,t) = \int_{-\infty}^{\infty} y^p c(y,r,t)dy \qquad \text{(B12)}$$

be the pth moment about the origin moving with speed V and

$$m^{(p)}(t) = \int_0^{r_1} 2rc_1^{(p)}(r,t) + \int_{r_1}^{r_2} 2rc_2^{(p)}(r,t)dr. \tag{B13}$$

Then

$$\frac{\partial c_i^{(p)}}{\partial t} = D_i \frac{1}{r}\frac{\partial}{\partial r}\left[r\frac{\partial c_i^{(p)}}{\partial r}\right] + D_i p(p-1)c_i^{(p-2)}$$
$$+ \{U_i\phi_i(r)-V\}pc_i^{(p-1)} \tag{B14}$$

with

$$\frac{\partial c_i^{(p)}}{\partial r} = 0, \ r=0; \quad \frac{\partial c_2^{(p)}}{\partial r} = 0, \ r=r_2 \tag{B15}$$

and

$$D_i \frac{\partial c_1^{(p)}}{\partial r} = D_2 \frac{\partial c_2^{(p)}}{\partial r} = k(c_2^{(p)}-\alpha c_1^{(p)}), \ r=r_1 \tag{B16}$$

Initially

$$c_i^{(p)}(r,0) = \chi_i^{(p)}(r) = \int_{-\infty}^{\infty} x^p\chi_i(x,r)dx \tag{B17}$$

and it is of course assumed that these moments are finite.

These equations come easily from multiplying the earlier equations by y^p and integrating from $-\infty$ to ∞. To get equations for $m^{(p)}$ we have to average over the cross-section, giving

$$\frac{dm^{(p)}}{dt} = p(p-1)\sum_1^2 D_i \int_{r_{i-1}}^{r_i} 2rc_i^{(p-2)}(r)dr + p\sum_1^2 \int_{r_{i-1}}^{r_i} 2r\{U_i\phi_i(r)-V\}\, c_i^{(p-1)}dr \tag{B18}$$

with

$$m^{(p)}(0) = \sum_1^2 \int_{r_{i-1}}^{r_i} 2r\chi_i^{(p)}(r)dr. \tag{B19}$$

It is perhaps well to interpolate here the results for Γ_4 since they can be misunderstood. Without loss of generality we can take the initial distribution to be such that $m^{(0)}(0)=1$, $m^{(1)}(0)=0$. It can then be shown

[11] that $m^{(0)}(t)=1$, $c_1^{(0)}(t)\to\pi/(A_1+\alpha A_2)$, $c_2^{(0)}(t)\to\alpha\pi/(A_1+\alpha A_2)$ where $A_i=\pi(r_i^2-r_{i-1}^2)$. Moreover if V is chosen to be $(A_1U_1+\alpha A_2U_2)/(A_1+\alpha A_2)$ then $dm^{(1)}/dt\to 0$. This means that the centre of gravity of the solute ultimately moves with the weighted mean speed of the two streams, the weights being the amounts of the solute in the two phases. This is very reasonable but it should be emphasized that it is an asymptotic result and, while one can give estimates of how quickly it is approached, the exact value of $m^{(1)}$ depends on the initial distribution. If $\beta_1=A_1/(A_1+\alpha A_2)\beta_2=1-\beta_1$, the asymptotic result for $m^{(2)}$ is that its growth rate

$$\frac{dm^{(2)}}{dt}\to 2\sum_1^2\beta_i\left[D_i+\kappa_i\left\{\frac{U_i^2(r_i^2-r_{i-1}^2)}{D_i}\right\}+\frac{\beta_1^2\beta_2^2(A_1+\alpha A_2)(U_1-U_2)^2}{\pi k\alpha r_1}\right] \tag{B20}$$

where

$$\kappa_i=\kappa_{i1}-2\kappa_{i2}(V/U_i)+\kappa_{i3}(V/U_i)^2 \tag{B21}$$

and the κ_{ij} can be calculated from a knowledge of the velocity profiles. For example, when $U_2=0$(hypothesis H_5) and $r_2\to r_1$ (H_6) the first two terms are $\beta_1[D_1+(U_1^2r_1^2/48D_1)(11-16\beta_1+6\beta_1^2)]$ and $\beta_2[D_2+(U_1^2(r_2-r_1)^2/3D_2)(1-2\beta_2+\beta_2^2)]$ respectively. Again these growth rates are only asymptotically valid and nothing more is rigorously claimed for this approach. It is quite another step to draw comparisons with apparent longitudinal dispersion coefficients such as is done in subsequent models.

The Taylor diffusion models with laminar flow.

We now specialize from the chromatographic case to the case to Taylor diffusion in laminar flow, i.e. we invoke H_7 and H_{14}. As before, we could let the system be much more general (cf. [10]) but this is not to the point here. We then have Δ_1 corresponding to the most complete equations for the physical system D. We can drop the suffixes since there is only one region and concentration and will write $r_1=a$. Then

$$\frac{\partial c}{\partial t} + 2U\left[1 - \frac{r^2}{a^2}\right]\frac{\partial c}{\partial x} = D\,\frac{1}{r}\frac{\partial}{\partial r}\left[r\,\frac{\partial c}{\partial r}\right] + D\,\frac{\partial^2 c}{\partial x^2} \tag{B22}$$

and

$$\frac{\partial c}{\partial r} = 0, \quad r = 0, a \tag{B23}$$

$$c(x,r,t) \text{ finite as } x\to\pm\infty \tag{B24}$$

$$c(x,r,0) = \chi(x,r) \tag{B25}$$

This is the system Δ_1 which can be modified to Δ_1' by the change of variable

$$y = x-Ut \tag{B26}$$

giving

$$\frac{\partial c}{\partial t} + U\left[1-2\,\frac{r^2}{a^2}\right]\frac{\partial c}{\partial y} = D\,\frac{1}{r}\frac{\partial}{\partial r}\left[r\,\frac{\partial c}{\partial r}\right] + D\,\frac{\partial^2 c}{\partial y^2}\,. \tag{B27}$$

Gill et al. [74] invoked H_2 and assumed that $c(o,r,t)$ could be specified, in fact as a constant which without loss of generality can be taken to be 1. Thus to (B22) and (B23) there is added

$$c(o,r,t) = 1, \tag{B28}$$

$$c(x,r,t)\to 0 \text{ as } x\to\infty, \tag{B29}$$

$$c(x,r,0) = 0 \tag{B30}$$

This is Δ_2 and is worth noting that Gill made them nondimensional in a way that would bring out the scale of the spread into which the sharp front at $t=0$, $x=0$ will soften as t increases. Thus with

$$\rho = \frac{r}{a},\ \xi = \frac{x}{2U}\sqrt{\frac{D}{t}},\ \tau = \frac{Dt}{a^2},\ P = \frac{2aU}{D} \tag{B31}$$

we have

$$\frac{\partial c}{\partial \tau} + \left[\frac{1-\rho^2}{\tau^{1/2}} - \frac{\xi}{2\tau}\right] \frac{\partial c}{\partial \xi} = \frac{\partial^2 c}{\partial \rho^2} + \frac{1}{\rho}\frac{\partial c}{\partial \rho} + \frac{1}{4\tau P^2}\frac{\partial^2 c}{\partial \xi^2} \tag{B32}$$

with

$$\frac{\partial c}{\partial \rho} = 0, \ \rho = 0,1, \tag{B33}$$

$$c(0,\rho,\tau) = 1, \tag{B34}$$

$$c(\xi,\rho,\tau) \to 0 \text{ as } \xi \to \infty. \tag{B35}$$

If H_{11} is invoked and we define

$$\begin{aligned} c^{(p)}(r,t) &= \int_{\infty}^{\infty} y^p c(y,r,t)dy \\ m^{(p)}(t) &= \frac{1}{a^2}\int_0^a 2rc^{(p)}(r,t)dr \end{aligned} \tag{B36}$$

then

$$\frac{\partial c^{(p)}}{\partial t} = D\frac{1}{r}\frac{\partial}{\partial r}\left[r\frac{\partial c^{(p)}}{\partial r}\right] + Dp(p-1)c^{(p-2)} + pU\left[1 - \frac{r^2}{a^2}\right]c^{(p-1)} \tag{B37}$$

and

$$\frac{dm^{(p)}}{dt} = Dp(p-1)m^{(p-2)} + p\frac{U}{a^2}\int_0^a 2r(1-\frac{r^2}{a^2})c^{(p-1)}(r,t)dr. \tag{B38}$$

If $m^{(0)}(0) = 1$ and the $m^{(p)}(0)$ are finite then $m^{(0)}(t) = 1$, $m^{(1)}(t) \to$ a constant and $dm^{(2)}(t)/dt \to 2[D+(a^2U^2/48D)]$. These are again asymptotic results for this model, Δ_3. The detailed asymptotic approach to normality of the distribution of solute has been discussed in an excellent paper by Chatwin [42].

Models with the mean concentration.

We now have the bolder hypothesis that the situation can be sufficiently represented in terms of plug flow and an effective longitudinal diffusion coefficient which govern the mean concentration

$$\bar{c}(x,t) = \frac{1}{a^2} \int_0^a 2rc\,(x,r,t)dr. \tag{B39}$$

Using Gill's boundary conditions we write

$$\frac{\partial \bar{c}}{\partial t} + U \frac{\partial \bar{c}}{\partial x} = D_e \frac{\partial^2 \bar{c}}{\partial x^2} \tag{B40}$$

with

$$\bar{c}(0,t) = 1 \tag{B41}$$

$$\bar{c}(x,t) \to 0 \text{ as } x \to \infty \tag{B42}$$

$$\bar{c}(x,0) = 0 \tag{B43}$$

This is Δ_4 and we observe that is is cognate with Δ_1 but not derivable from it, for averaging over the cross-section gives

$$\frac{\partial \bar{c}}{\partial t} + U \frac{\partial \langle c \rangle}{\partial x} = D \frac{\partial^2 \bar{c}}{\partial x^2}$$

where

$$\langle c \rangle = \frac{4}{a^2} \int_0^a r \left[1 - \frac{r^2}{a^2} \right] c(x,r,t)dr \tag{B44}$$

is not $\bar{c}$, but the cup-mixing mean. By subtraction

$$D_e = D + U \frac{\partial}{\partial x} [\bar{c} - \langle c \rangle] \Big/ \frac{\partial^2 \bar{c}}{\partial x^2} \tag{B45}$$

and this equation has been used to calculate a variable D_e from any particular solution.

The model Δ_4 may be varied by the choice of D_e and we recognize the following possibilities. If H_{13} is invoked then Taylor [174,175] showed that

$$\Delta_4' : D_e = \frac{a^2 U^2}{48D} \tag{B46}$$

The growth of moments in Δ_4 and Δ_3 are asymptotically the same if

$$\Delta_4' : \quad D_e = D + \frac{a^2U^2}{48D} \tag{B47}$$

On the other hand diffusion may dominate completely in a very slow flow and

$$\Delta_4''' : \quad D_e = D. \tag{B48}$$

Clearly Δ_4' and Δ_4''' are limiting cases of Δ_4''.

At the other end of the spectrum we might assert diffusion is altogether negligible, H_{15}. In this case the flat profile at $t=0$ becomes a paraboloid whose tip advances with twice the mean speed. For this

$$\Delta_5 \quad \bar{c} = \begin{cases} 1-x/2Ut, & x<2Ut, \\ 0, & x>2Ut. \end{cases} \tag{B49}$$

<u>The correlation of results from a model.</u>

We have already mentioned that an "exact apparent" diffusion coefficient can be fitted to the results of the calculation with Δ_2 so that in Δ_4 it reproduces the same mean concentration. Unless some serendipitous constancy emerges this exercise serves only to compare the models for the particular problem. Another method is to look for a correlation between the results of one model and those of another without attempting to give significance to the constants. This is again limited in usefulness to a particular problem and is an empirical exercise with interpolatory value but no extrapolatory or explanatory power. For example Gill et al. [74] found that a good approximation to the solution of Δ_2 was afforded by the formula

$$\bar{c} = \frac{1}{2}\,\text{erfc}\,\frac{x-Ut}{2\sqrt{Dt(1+P^2Q)}} + \frac{1}{2}\exp\frac{xU}{D(1+P^2Q)}\,\text{erfc}\,\frac{x+Ut}{2\sqrt{Dt(1+P^2Q)}} \tag{B50}$$

where, as before, $P = aU/D$ and

$$\Delta_6: \quad Q = 0.028(Dt/a^2)^{0.55}, \quad Dt/a^2 < 0.6 \tag{B51}$$

It is worth noting that Δ_4'' corresponds to Δ_5 but with $Q = 1/48$, while Δ_4''' would correspond to $Q = 0$.

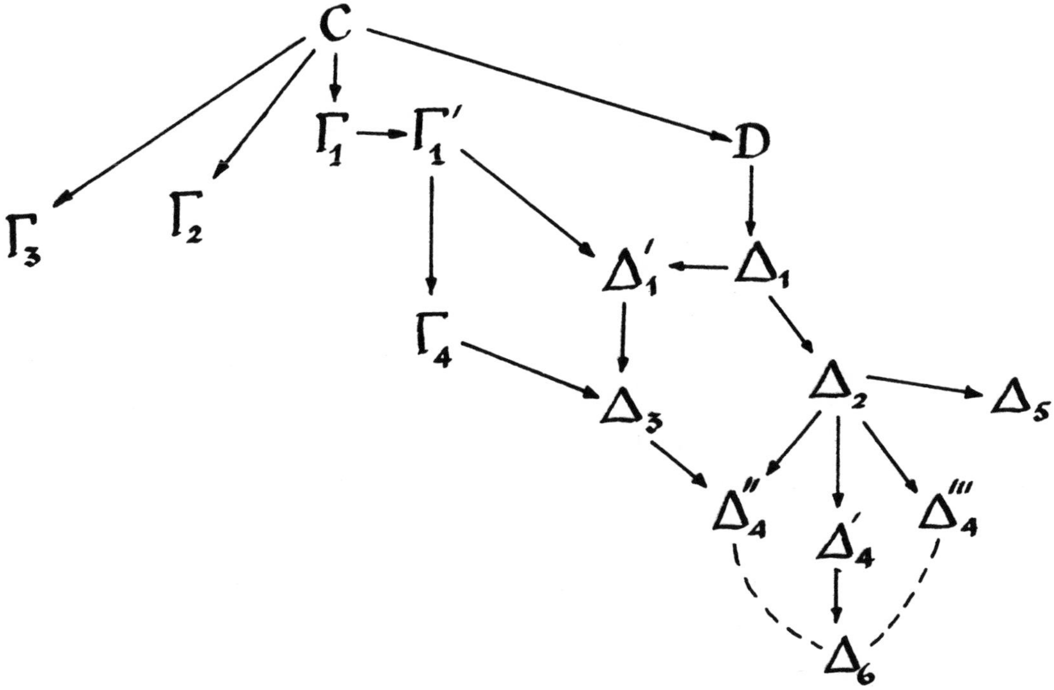

Summary.

Model	Hypotheses	Equations
Γ_1	0,1,4,8,9,10	B1 - B6
Γ_1'	0,1,4,8,9,10	B1 - B8
Γ_2	0,2,4,8,9,10	B1 - B5,B9
Γ_3	0,3,4,8,9,10	B1 - B5, B10,B11
Γ_4	0,1,4,8,9,10,11	B12 - B19
Δ_1	0,1,4,7,9,10,14	B22 - B25
Δ_1'	0,1,4,7,9,10,14	B23 - B27
Δ_2	0,2,4,7,9,10,14	B22,B23,B28 - B30
Δ_2'	0,2,4,7,9,10,14	B31 - B35
Δ_3	0,2,4,7,9,10,11,14	B36 - B38
Δ_4	0,2,4,7,9,10,12,14	B39 - B43
Δ_5	0,2,4,7,9,10,14,15	B49
Δ_6	0,2,4,7,9,10,12,14	B50 - B51

Notation.

A_i	area of ith phase $\pi(r_i^2 - r_{i-1}^2)$
a	radius of tube, common value of r_1 and r_2
c	concentration of solute in tube (system D)
c_i	concentration of solute in i^{th} phase
c_{if}	feed concentration
$c_i^{(p)}, c^{(p)}$	p^{th} moment of concentration
$\bar{c}$	average concentration
$\langle c \rangle$	cup mixing mean concentration
D_e	equivalent longitudinal dispersion coefficient

D_i	diffusion coefficient in i^{th} phase
k	rate constant for partition
L	length of tube
$m^{(p)}$	p^{th} moment of mean concentration
P	Peclet number aU/D
Q	defined in eqn. (B51)
r	radial distance (r_o=0; r_1 interface of phases; r_2, outer radius of phase 2)
t	time
U_i	mean velocity of phase i
V	velocity of moving origin
x	length coordinate
y	x-Vt
α	partition coefficient
β_i	$A_i/(A_1 + A_2)$
κ	constants in eqn. (B21)
ξ	$xD^{1/2}/2Ut^{1/2}$
ρ	r/a
τ	Dt/a^2
ϕ_i	velocity distribution
χ_i	initial distribution
$\chi_i^{(p)}$	p^{th} moment of initial distribution

Appendix C The stirred tank reactor

Description of the system S.

A stirred tank reactor consists of a cylindrical vessel of volume V with incoming and outgoing pipes. The incoming pipes bring reactants $A_1, A_2 \ldots A_r$, at volume flow rates $q_1, q_2, \ldots q_r$, and the outgoing pipe takes of the mixture of products $A_{r+1}, \ldots A_s$ and the remnants of the reactants, at a flow rate of $q = q_1 + q_2 + \ldots + q_r$. Thus the volume V remains constant. The reaction can be written as $\Sigma\alpha_j A_j = 0$, where $\alpha_{r+1}, \ldots \alpha_s$ are

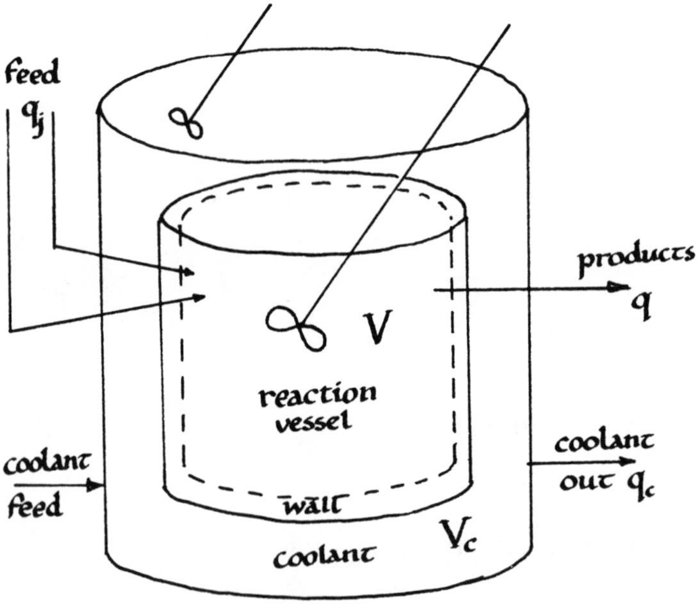

positive. This cylinder is immersed in another cylinder of annular volume V_c, also perfectly stirred, which is fed cooling water of temperature T_{cf} at a flow rate of q_c. Other details will be mentioned as we proceed. It should be mentioned that this description has already been deliberately simplified since the geometry of a real jacketed reactor would undoubtedly be more complicated than that of simple cylinders. However, I have no desire to pile Pelion on Ossa.

Hypotheses.

Let us lump together the applicability of all physical laws, such as the conservation of matter and energy or Fourier's law of heat conduction and call this the underlying general hypothesis, H_o. The following hypotheses can be extracted from the description or be excogitated as relevant to the setting:

H_1: the mixing is perfect so that the concentrations c_j, the reaction temperature T and T_c, the temperature of the coolant jacket, are all independent of position, though they may be functions of time. The volumes V and V_c are constant, as also are the flow rates q_j and the feed temperatures T_{jf}. The work done by the stirrers may be ignored.

H_2: the reaction rate is a function $r(c_1,\ldots c_s,T)$ such that the rate of change in the number of moles of A_j by reaction alone is $\alpha_j r$ per unit volume.

H_3: the heat transfer to the inner and outer sides of the wall where the surface temperatures will be denoted by T_i and T_o respectively can be described by transfer coefficients h_i and h_o such that the heat transferred per unit area is $h_i(T-T_i)$ and $h_o(T_o-T_c)$ respectively.

H_4: the heat capacity of the reaction mixture does not change significantly.

H_5: the system is in steady state.

H_6: the curvature of the wall is negligible and the sharp corners can be ignored.

H_7: the conductivity of the wall is extremely high.

H_8: the heat capacity of the wall is negligible.

H_9: the response of the cooling jacket is virtually instantaneous.

H_{10}: the reaction is the first-order and irreversible with respect to the key species.

Derivation of the most general model Σ_1.

Using the principle of the conservation of matter from the underlying hypothesis H_o, we have the following balance for each species:

$$\begin{bmatrix}\text{rate of change of}\\ \text{number of moles of}\\ A_j \text{ in reactor}\end{bmatrix} = \begin{bmatrix}\text{rate of}\\ \text{feed of}\\ A_j\end{bmatrix} - \begin{bmatrix}\text{rate of}\\ \text{withdrawal}\\ \text{of } A_j\end{bmatrix} + \begin{bmatrix}\text{rate of formation}\\ \text{of } A_j \text{ by means of}\\ \text{the reaction}\end{bmatrix}$$

If c_{jf} is the concentration (moles/volume) of A_j in its feed stream, this translates immediately into the ordinary differential equation

$$V \frac{dc_j}{dt} = q_j c_{jf} - qc_j + \alpha_j Vr(c_1,\ldots c_S,T). \tag{C1}$$

In obtaining this equation we have invoked H_1 and H_2.

If $h_j(c_1,\ldots c_S,T)$ is the enthalpy per mole of A_j, and the work done by the stirrer is ignored, then conservation of energy implies

$$V \frac{d}{dt} \Sigma c_j h_j = \Sigma q_j c_f h_{jf} - q\Sigma c_j h_j - A_i h_i (T-\bar{T}_i)$$

The symbol h_{jf} denotes the specific enthalpy of A_j evaluated for its feed conditions. In the last term A_i is the total internal wall area and since the heat transfer coefficient h_i is independent of position we need only average the inner surface temperature of the wall. Thus H_1 and H_3 are used here. We now simplify this equation by subtracting from it the sum over j of equations (C1) each multiplied by h_j. Thus

$$V\Sigma c_j \frac{dh_j}{dt} = q_j c_{jf}(h_{jf}-h_j) - (\Sigma\alpha_j h_j)V_r - A_i h_i(T-\bar{T}_i)$$

Next we observe that $\alpha_j h_j = \Delta H$ is the heat of reaction and that since h_j is an intensive thermodynamic variable (i.e. $\Sigma c_j \frac{\partial h_j}{\partial c_k} = \Sigma c_j \frac{\partial h_k}{\partial c_j} = 0$)

$$\Sigma c_j \frac{dh_j}{dt} = \Sigma c_j c_{pj} \frac{dT}{dt} + \Sigma\Sigma c_j \frac{\partial h_j}{\partial c_k} \frac{dc_k}{dt} = C_p \frac{dT}{dt}$$

where c_{pj} is the heat capacity of A per mole and C_p is the heat capacity of the mixture per unit volume. We now invoke H_4 and write $q_j c_{jf}(h_{jf}-h_j) = qC_p(T_f-T)$ to give

$$VC_p \frac{dT}{dt} = qC_p(T_f-T) + (-\Delta H)Vr(c_1,\ldots T) - A_i h_i(T-\bar{T}_i). \tag{C2}$$

This form allows us to check the common sense of the equation for we can write it as:

$$\begin{bmatrix}\text{rate of}\\ \text{change}\\ \text{of heat}\\ \text{content}\end{bmatrix} = \begin{bmatrix}\text{heat}\\ \text{brought}\\ \text{in with}\\ \text{feed}\end{bmatrix} - \begin{bmatrix}\text{heat}\\ \text{taken out}\\ \text{with}\\ \text{products}\end{bmatrix} + \begin{bmatrix}\text{heat}\\ \text{generated}\\ \text{by}\\ \text{reaction}\end{bmatrix} - \begin{bmatrix}\text{heat}\\ \text{removed by}\\ \text{cooling}\\ \text{wall}\end{bmatrix}$$

The wall has been simplified to be a finite cylinder of internal area, A_i. If we denote the region it occupies by D and its inner and outer surfaces by ∂D_i and ∂D_o, we apply conservation principles and Fourier's law of heat conduction to obtain for the wall temperature T_w,

$$\rho_w c_{pw} \frac{\partial T_w}{\partial t} = k_w \nabla^2 T_w \text{ in } D, \tag{C3}$$

where ρ_w, c_{pw} and k_w are the density, specific heat and conductivity of the wall respectively. To obtain boundary conditions we have to call on H_3,

$$k_w \frac{\partial T_w}{\partial n} = h_i(T-T_i) \text{ on } \partial D_i, \tag{C4}$$

$$k_w \frac{\partial T_w}{\partial n} = h_o(T_c - T_o) \text{ on } \partial D_o, \tag{C5}$$

where $\partial/\partial n$ is the normal derivative on the surface directed outward from D.

Before seeing how these equations simplify let us write the heat balance equation for the coolant. This is

$$V_c C_{pc} \frac{dT_c}{dt} = q_c C_{pc}(T_{cf} - T_c) + A_o h_o(\bar{T}_o - T_c) \tag{C6}$$

where C_{pc} is the heat capacity of the coolant per unit volume, A_o is the area of ∂D_o and $\bar{T}_o$ the average outer temperature. H_1 and H_3 have been involved in deriving this equation as well as the underlying H_o.

Equations (C1-6), together with suitable initial conditions, give $s+2$ ordinary and one partial differential equation with its boundary conditions and constitute Σ_1, the most detailed model we shall consider. In obtaining it the hypothesis used have been H_o, H_1, H_2, H_3 and H_4.

Derivation of the steady state models Σ_2 and Σ_3.

Now let us invoke H_5 and assume that the system is at steady state. To do this is to set all time derivatives equal to zero and it leads to a partial differential equation, Laplace's, for T_w connected through its boundary conditions to a set of algebraic equations. (The term algebraic equation is applied to any equation that is not a differential equation even though transcendental functions may appear in it.) Let this model be Σ_2.

However when we recognize that T_w is a potential function we can use Green's theorem to give

$$\begin{aligned} 0 &= \iiint_D k_w \nabla^2 T_w dV = \iint_{\partial D_i + \partial D_o} k_w(\partial T_w/\partial n) dS \\ &= A_i h_i(T - \bar{T}_i) - A_o h_o(\bar{T}_o - T_c) \end{aligned} \tag{C7}$$

Combining this with (C6) we have three expressions for the rate of removal of heat

$$Q_c = q_c C_{pc}(T_c - T_{cf}) = A_o h_o(\bar{T}_o - T_c) = A_i h_i(T - \bar{T}_i).$$

This gives

$$T - T_{cf} = Q_c \left\{ \frac{1}{q_c C_{pc}} + \frac{1}{A_o h_o} + \frac{1}{A_i h_i} + \frac{\bar{T}_i - \bar{T}_o}{Q_c} \right\}$$

but the last term is a messy one and has to be evaluated from the full solution of the potential equation. If however we invoke H_6 and take d_w to be the thickness of the wall then we have the local flux of heat per unit area equal to $k_w(T_i - T_o)/d_w$. Also ignoring curvature makes $A_o = A_i = A$ so that $Q_c = k_w A(T_i - T_o)/d_w$. Then

$$Q_c = \hat{h}A(T - T_{cf}) \text{ where } \frac{1}{\hat{h}} = \frac{A}{q_c C_{pc}} + \frac{1}{h_o} + \frac{1}{h_i} + \frac{d_w}{k_w} \qquad \text{(C8)}$$

Thus the model simplifies to Σ_3 consisting of the equations

$$q_j c_{jf} - q c_j + \alpha_j V r(c_1, \ldots c_s, T) = 0, \; j = 1, \ldots s \qquad \text{(C9)}$$

$$qC_p(T_f - T) + (-\Delta H)Vr(C_1, \ldots c_s, T) - \hat{h}A(T - T_{cf}) = 0 \qquad \text{(C10)}$$

Notice that H_8 has no relevance at all to this model and that were we to invoke H_7 it would not change the model but only modify the value of $\hat{h}$.

<u>Simplified transient models Σ_4 and Σ_5.</u>

Let us return now to the transient model, dropping H_5, and see the effect of H_7 and H_8. A little caution is needed here or terms can get lost. First suppose that $k_w \to \infty$ (H_7), we cannot conclude that T_w is constant in time by observing that this limit, like the steady state hypothesis, leads to Laplace's equation for T_w. Rather the form of the temperature profile through the wall is invariant in time, for it takes up a temperature profile between T_i and T_o with virtually no delay. Thus, for very large k_w, the Laplacian of the temperature in the wall becomes very small and the product $k_w \nabla^2 T_w$ is finite.

Let T_w denote the mean temperature of the wall, which is uniform with respect to position around the reactor on account of $k_w \to \infty$ and the fact that the wall is exposed to uniform temperatures on both sides. Then integrating (C3) throughout D, using Green's theorem and the boundary conditions (C4) and (C5), gives

$$V_w \rho_w C_{pw} \frac{dT_w}{dt} = A_i h_i T + A_o h_o T_c - (A_i h_i + A_o h_o) T_w \tag{C11}$$

Thus we have a model consisting of (S+3) ordinary differential equations (C1), (C2), (C6) and (C11), where in (C2) and (C6), T_i and T_o have both been made equal to T_w. Let this be the model Σ_4; it invokes H_o-H_4 and H_7. Note that we did not need H_8.

Suppose now that we assert H_8, but drop H_7. Again T_w is governed by Laplace's equation and we can arrive at (C7). However we are still left with Laplace's equation unless we invoke H_6 and note that

$$Q_c = A\ h_o(\bar{T}_o - T_c) = Ak_w(\bar{T}_i - \bar{T}_o)/d_w = Ah_i(T - \bar{T}_i) = Ah^*(T - T_c) \tag{C12}$$

where

$$\frac{1}{h^*} = \frac{1}{h_o} + \frac{d_w}{k_w} + \frac{1}{h_i}$$

We then have a model Σ_5, consisting of (S+2) ordinary equations; the S equation (C1) and

$$VC_p \frac{dT}{dt} = qC_p(T_f - T) + (-\Delta H)Vr(c_1, \ldots T) - Ah^*(T - T_c) \tag{C13}$$

$$V_c C_{pc} \frac{dT_c}{dt} = q_c C_{pc}(T_{cf} - T_c) + Ah^*(T - T_c) \tag{C14}$$

Finally two more hypotheses are introduced to give a specially important case. The first, H_9, is of the nature of the limiting hypotheses we have been making. It asserts that $V_c/q_c << V/q$ so that, by comparison with the other time derivatives, dT_c/dt has a small multiplier. If we go to the limit and say that the response of the cooling jacket is virtually instantaneous, we wipe out the derivative in eqn. (C6). Thus we are essentially back in the steady state and can write $Q_c = hA(T-T_{cf})$ as in eqn. (C8).

The second hypothesis, H_{10}, asserts that the reaction is irreversible and first order in the concentration of one species. If this is A_1 we can write $\alpha_1 = -1$ and $r = kc_1$, where $k = k(T)$ is a function only of T. For simplicity we can then drop the suffix on c and eqns. (C1) and (C2) become

$$V \frac{dc}{dt} = q(c_f-c) - Vk(T)c \tag{C15}$$

$$VC_p \frac{dT}{dt} = qC_p(T_f-T) + (-\Delta H)Vk(T)c - hA(T-T_{cf}) \tag{C16}$$

These equations give a pair of equations for a pair of unknowns $c(t)$, $T(t)$.

<u>The dimensionless equations</u>.

Up to this point everything has been very dimensional and it is not clear what we have mean by large and small values. There are various characteristic lengths, times etc. in the problem and we want to pick the most judicious set. In particular constants to which we are going to give some limiting value should not be used to render others dimensionless, nor should those whose variation we are going to study.

Let V/q be the characteristic time and $\tau = qt/V$;

c^*, be a characteristic concentration, say $\Sigma q_j c_{jf}/q$ and $u_j = c_j/c^*$;

T^*, be a characteristic temperature, say $(-\Delta H)c^*/C_p$, and $v = T/T^*$, $w = T_w/T^*$, $\Theta = T_c/T^*$, $v_f = T_f/T^*$, $\Theta_f = T_{cf}/T^*$, $w_i = T_i/T^*$, $w_o = T_o/T^*$;

d_w, be a characteristic length such as a mean wall thickness with which the independent variables in the Laplacian are to be made dimensionless.

The other parameters will emerge with the equations. If (C1) is divided by qc^* it becomes

$$\frac{du_j}{d\tau} = \gamma_j - u_j + \alpha_j R(u_1,\ldots,v) \qquad \text{(C18)}$$

where $\gamma_j = q_j c_{jf}/\Sigma q_j c_{jf}$, i.e. $\Sigma\gamma_j = 1$, is the jth fraction of feed and $R = Vr/qc^*$. Similarly let (C2) be divided by qC_pT^* to give

$$\frac{dv}{d\tau} = v_f - v + R(u_1,\ldots v) - \beta_i(v - v_i) \qquad \text{(C19)}$$

where $\beta_i = A_i h_i/qC_p$.

We will use the same symbol as before for the Laplacian with respect to the dimensionless variables, so that (C3) can be divided by $\rho_w c_{pw} q/VT^*$ to give

$$\frac{\partial w}{\partial \tau} = \lambda \nabla^2 w \qquad \text{(C20)}$$

where $\lambda = k_w V/q\rho_w c_{pw} d_w^2$. If $\partial/\partial\nu$ denotes the normal derivative in the dimensionless variables

$$\lambda \frac{\partial w}{\partial \nu} = \frac{\beta_i}{\delta_w}(v - w_i) \text{ on } \partial D_i \qquad \text{(C21)}$$

where $\delta_w = A_i d_w \rho_w c_{pw}/VC_p$ is the ratio of the heat capacity of the wall to that of the contents. Similarly

$$\lambda \frac{\partial w}{\partial \nu} = \frac{\beta_o}{\delta'_w} (\Theta - w_o) \text{ on } \partial D_o \tag{C22}$$

where $\beta_o = A_o h_o / qC_p$ and $\delta'_w = (A_o/A_i)\delta_w$. Finally (C5) becomes

$$\delta_c \frac{d\Theta}{d\tau} = \chi(\Theta_f - \Theta) + \beta_o(w_o - \Theta) \tag{C23}$$

where $\delta_c = V_c C_{pc}/VC_p$ and $\chi = q_c C_{pc}/qC_p$.

In the later models we have $\hat{h}$ and h^* and we make them dimensionless with qC_p to give

$$\hat{\beta} = A\hat{h}/qC_p = \left\{\frac{1}{\chi} + \frac{1}{\beta_o} + \frac{1}{\beta_i} + \frac{1}{\lambda\delta_w}\right\}^{-1} \tag{C24}$$

$$\beta^* = Ah^*/qC_p = \left\{\frac{1}{\beta_o} + \frac{1}{\lambda\delta_w} + \frac{1}{\beta_i}\right\}^{-1} \tag{C25}$$

Thus we have for Σ_3 the (S+1) non-differential equations

$$\gamma_j - u_j + \alpha_j R(u_1, \ldots v) = 0, \tag{C26}$$

$$v_f - v + R(u_1, \ldots v) - \hat{\beta}(v - \Theta_f) = 0 \tag{C27}$$

The model Σ_4 consists in eqns. (C18), (C19) and (C23) with the dimensionless form of (C11), namely

$$\delta_w \frac{dw}{d\tau} = \beta_i v + \beta_o \Theta - (\beta_o + \beta_i) w. \tag{C28}$$

Finally, the model Σ_5 in eqn. (C15) and the two equations

$$\frac{dv}{d\tau} = v_f - v + R(u_1, \ldots v) - \beta^*(v - \Theta) \tag{C29}$$

$$\delta_c \frac{d\Theta}{d\tau} = \chi(\Theta_f - \Theta) + \beta^*(v - \Theta) \tag{C30}$$

The initial conditions, as needed, are

$$u_j = u_{jo}, v = v_o, \; w = w_o(\underset{\sim}{\xi}), \; \Theta = \Theta_o, \; \tau = 0, \; \underset{\sim}{\xi} = \underset{\sim}{x}/d_w \tag{C31}$$

The ways in which Σ_6 can be non-dimensionalized are discussed extensively in Sec. 4.2.

Summary of parameters.

Reaction:	α_j	stoichiometric coefficients
	—	parameters of the rate law e.g. E/RT_f
Feed:	γ_j	fraction of A_j in feed
	v_f	feed temperature
	Θ_f	coolant feed temperature
Capacities:	δ_c	heat capacity ratio of coolant to reactants
	δ_w	ratio of heat capacity of wall to reactants
	χ	ratio of heat carrying capacities of coolants to reactants
Transfer:	β_o, β_i	dimensionless heat transfer coefficients
	$\hat{\beta}, \beta^*$	composite heat transfer-coefficients
	λ	dimensionless wall conductivity

Summary of models.

Model Σ	Hypotheses H	Equations C	Dimensionless equations	Remarks
1	0,1,2,3,4	1,2,3,4,5,6	18,19,20,21,22,23	
2	0,1,2,3,4,5	1,2,3,4,5,6	18,19,20,21,22,23	Set $\partial/\partial t$ or $\partial/\partial\tau=0$
3	0,1,2,3,4,5,6	9,10	26,27	
4	0,1,2,3,4,7	1,2,6,11	18,19,23,28	
5	0,1,2,3,4,6,8	1,13,14	18,29,30	
6	0,1,2,3,4,6,8,9,10	15,16	See Sec. 4.2	

Notation for Appendix C: the system S and its models Σ.

A_j	chemical species, j = 1, S; j = 1...r for reactants, r+1,...S products
A_i, A_o	inner and outer areas of reactor wall
C_p	heat capacity per unit volume of reaction mixture

C_{pc}	heat capacity per unit volume of coolant
c_j	concentration of A_j
c_{jo}	initial concentration of A_j
c_{jf}	feed concentration of A_j
c_{pw}	specific heat of wall
c^*	reference concentration
d_w	thickness of reactor wall
h_j	enthalpy per mole of A_j
h_{jf}	enthalpy per mole of A_j under feed conditions
h_i, h_o	heat transfer coefficient at inner and outer wall surfaces
$\hat{h}, h^*$	composite heat transfer coefficients
k_w	thermal conductivity of wall
n	outward normal to wall in $\partial/\partial n$
Q_c	total rate of heat removal
q	flow rate of reacting mixture
q_c	coolant rate flow
q_j	feed rate of A_j
R	dimensionless reaction rate Vr/qc^*
r	reaction rate per unit volume
S	number of reacting species
T	temperature
$T_c, T_{cf}, T_f, T_w(\underset{\sim}{x})$	temperature of coolant, coolant feed, reactor feed and wall resp.
$T_i, T_o, \bar{T}_i, \bar{T}_o$	inner and outer wall temperatures and their averages
T_o', T_{wo}, T_o	initial reactor, wall and coolant temperatures
T^*	reference temperature
t	time

V	volume of reactor
V_c, V_w	volume of coolant, wall
v	dimensionless temperature T/T^*
v_f	T_f/R^*
$w(\xi)$	T_w/T^*
w_i/w_o	$T_i/T^*, T_o/T^*$
$\underset{\sim}{x}$	coordinates within the wall
α_j	stoichiometric coefficients
$\hat{\beta}_i, \beta_o$	$h_i A_i/qC_p, h_o A_o/qC_p$
β, β^*	dimensionless composite heat transfer coefficients; (C24), (C25)
γ_j	dimensionless feed rate of A_j
$\delta_c, \delta_w, \delta'_w$	$V_c C_{pc}/VC_p, A_i d_w \rho_w c_{pw}/VC, A_o d_w \rho_w c_{pw}/VC_p$
ΔH	heat of reaction
Θ	dimensionless coolant temperature, T_c/T^*
Θ_o, Θ_f	$T_{co}/T^*, T_{cf}/T^*$
λ	$k_w V/q_w c_{pw} d_w^2$
ν	dimensionless normal in $\partial/\partial\nu$
$\underset{\sim}{\xi}$	x/d_w
ρ_w	density of wall
τ	qt/V
χ	$q_c C_{pc}/qC_p$

References

1 P. Achinstein — Models, analogies and theories. Phil. of Sci. 31 (1964) 328.

2 P. Achinstein — Theoretical models. Brit. Jnl. for the Phil. of Sci. 16 (1965) 102.

3 R. Ackermann — Confirmatory models of theories. Brit. Jnl. for the Phil. of Sci. 16 (1966) 312.

4 W. P. Alston — Philosophy of language. (Englewood Cliffs: Prentice Hall, 1964).

5 N. R. Amundson and R. Aris — Some remarks on longitudinal diffusion or mixing in fixed beds. A.I.Ch.E. J. 3 (1957) 280.

6 N. R. Amundson and S-L Liu — Stability of adiabatic packed bed reactors. A simplified treatment. I.E.C. Fundamentals 1 (1962) 200.

7 L. Apostel — Towards the formal study of models in the non-formal sciences. In [68] p. 1.

8 M. A. Arbib — Theories of abstract automata. (Englewood Cliffs: Prentice Hall, 1968).

9 M. A. Arbib and E. G. Manes — The categorical imperative: arrows, structures and functors. (New York: Academic Press, 1975).

10 R. Aris — On the dispersion of a solute in a fluid flowing through a tube. Proc. Roy. Soc. A235 (1956) 67.

11 R. Aris — On the dispersion of a solute by diffusion, convection and exchange between phases. Proc. Roy. Soc. A252 (1959) 538.

12 R. Aris — Some problems in the analysis of transient behavior and stability of chemical reactors. Adv. in Chem. 109 (1972) 578.

13 R. Aris — On the ostensible steady state of a dynamical system. Rend. Lincei. Sr. VIII 57 (1974) 1.

14 R. Aris — The mathematical theory of diffusion and reaction in permeable catalysts. (Oxford: Clarendon Press, 1975) 2 vols.

15	R. Aris	How to get the most out of an equation without really trying. Chem. Eng. Educ. 10 (1976) 114.
16	R. Aris and A. E. Humphrey	The dynamics of a chemostat in which two organisms compete for a common substrate. Biotech. and Bioeng. 19 (1977) 1375.
17	R. Aris and D. L. Schruben	Transients in distributed chemical reactors. Part I. A simplified model. Chem. Eng. J. 2 (1972) 179.
18	R. Aris and S. Viswanathan	An analysis of the countercurrent moving bed reactor. SIAM/AMS Proceedings 8 (1974) 99.
19	A. M. Arthurs	Complimentary variational principles. (Oxford: Clarendon Press, 1970).
20	P. Auger	Models in science. Diogenes 52 (1965) 1.
21	J. E. Bailey	Lumping analysis of reactions in continuous mixtures. Chem. Eng. J. 3 (1972) 52.
22	M. S. Bartlett	Introduction to stochastic processes. (Cambridge: Cambridge Univ. Press, 1966).
23	J. W. L. Beament (Ed.)	Models and analogues in biology (Symposia of the Society for Experimental Biology No. 14). (Cambridge: Cambridge Univ. Press, 1960).
24	R. Bellman, K. L. Cooke and J.	A. Lockett, Algorithms, graphs and computers. (New York: Academic Press, 1970).
25	R. Bellman and M. Giertz	On the analytic formalism of the theory of fuzzy sets. Information Sciences 5 (1973) 149.
26	R. Bellman and L. A. Zadeh	Decision-making in a fuzzy environment. Management Sciences 17 (1970) 8141.
27	C. Berge	The theory of graphs and its applications. (London: Methuen, 1964).
28	C. Berge and A. Ghouila-Houri	Programming, games and transportation networks. (London: Methuen, 1965).
29	J. M. Beshers	Models and theory construction. In M. L. Barron (ed.)(1966) 590.

30 K. B. Bischoff — An extension of the general criterion for the importance of pore diffusion. Chem. Eng. Sci. 22 (1967) 525.

31 J. T. Bonner — Analogies in biology. Synthese 15 (1963) 275.

32 K. C. Bowen — Mathematical battles. Bull. I.M.A. 9 (1973) 310.

33 R. B. Braithwaite — Scientific explanation. A study of the function of theory, probability and law in science. (Cambridge: Cambridge Univ. Press, 1953).

34 R. B. Braithwaite — Models in the empirical sciences. In Nagel, E. et al. (ed.) (1962) 224.

35 M. Brodbeck — Models, meanings and theories. In [77] (1959) 373.

36 M. Brodbeck — Models, meaning and theories. In Symposium on Sociological Theory (ed. L. Gross)(New York: Harper and Row, 1959). (Also in Readings in the philosophy of the social sciences (ed. M. Brodbeck)(New York: Macmillan, 1968).

37 M. Bunge — Models in theoretical science. Akten des XIV Int. Kong. für Philosophie (Herber, Wien)(1968) 208.

38 R. R. Bush and F. Mosteller — A comparison of eight models. In R. R. Bush and W. K. Estes (eds.) Studies in mathematical learning theory. (Stanford: Stanford Univ. Press, 1961). Also in [111].

39 K. V. Bury — Statistical models in applied science. (New York: John Wiley, 1975).

40 H. Byerly — Model structures and model objects. Brit. Jnl. for the Phil. of Sci. 20 (1969) 135.

41 N. R. Campbell — The foundations of science; the philosophy of theory and experiment. (New York: Dover, 1957).

42 P. C. Chatwin — The approach to normality of the concentration distribution of a solute in a solvent flowing along a straight pipe. J. Fluid Mech. 43 (1970) 321.

43	R. F. Churchhouse	Discoveries in number theory aided by computers. Bull. I.M.A. 9 (1973) 15.
44	W. A. Coppel	Stability and asymptotic behavior of differential equations. (Boston: Heath, 1965).
45	C. A. Coulson	Mathematics and the real world. Bull. I.M.A. 9 (1973) 2.
46	C. A. Coulson	The rôle of mathematics in chemistry. Bull. I.M.A. 9 (1973) 206.
47	C. A. Coulson	Mathematical models. Bull. I.M.A. 10 (1974) 340.
48	D. R. Cox and H. O. Miller	The theory of stochastic processes. (New York: John Wiley, 1965).
49	J. Crank	Diffusion mathematics in medicine and biology. Bull. I.M.A. 12 (1976) 106.
50	J. Crank and R. D. Prahle	Melting ice by the isotherm migration method. Bull. I.M.A. 9 (1973) 12.
51	I. Dambska	Modèle et objet de la connaissance. Revue Intnl. de Phil 87 (1969) 34.
52	G. B. Dantzig	Linear programming and extensions. (Princeton: Princeton Univ. Press, 1963).
53	H. A. Deans and L. Lapidus	A computational model for predicting and correlating the behavior of fixed bed reactors. A.I.Ch.E. J. 6 (1960) 656.
54	D. J. De Solla Price	Automata and the origins of mechanism and mechanistic philosophy. Tech. and Culture 5 (1964) 5.
55	K. W. Deutsch	Mechanism, organism and society: some models in natural and social science. Phil. of Sci. 18 (1951) 230.
56	P. Duhem	The aim and structure of physical theory. (Trs. P. P. Wiener). (1st Edn. 1906). (Princeton: Princeton Univ. Press, 1954).
57	A. W. F. Edwards	Models in genetics. In [23] p. 6.
58	P. Edwards (ed.)	Encyclopedia of Philosophy. (London, 1967). (8 vols.)

59	G. Eigenberger	On the dynamic behavior of the catalytic fixed-bed reactor in the region of multiple steady states--I. The influence of heat conduction in two phase models. Chem. Eng. Sci. 27 (1972) 1909.
60	G. Eigenberger	On the dynamic behavior of the catalytic fixed-bed reactor in the region of multiple steady states--II. The influence of the boundary conditions in the catalyst phase. Chem. Eng. Sci. 27 (1972) 1917.
61	G. Eigenberger	Kinetic instabilities in catalytic reactions--a modelling approach. Proc. 4th Int. Symp. on Chem. Reaction Eng. Heidelberg, April, 1976. Dechema Frankfurt.
62	S. Eilenberg	Automata, languages and machines. (New York: Academic Press, 1974).
63	G. L. Farre	Remarks on Swanson's theory of models. Brit. Jnl. for the Phil. of Sci. 18 (1967) 140.
64	W. Feller	Introduction to probability theory and its applications (New York: John Wiley, 1968) 2 vols.
65	P. C. Fife	Pattern formation in reacting and diffusing systems. J. Chem. Phys. 64 (1976) 554.
66	B. A. Finlayson	The method of weighted residuals and variational principles. (New York: Academic Press, 1972).
67	L. Ford and D. Fulkerson	Flows in networks. (Princeton: Princeton Univ. Press, 1962).
68	H. Freudenthal	The concept and the role of the model in mathematics and social sciences. (Dordrecht: Reidel Pub. Co., 1961).
69	A. Friedman	Differential games. (New York: Wiley-Interscience, 1971).
70	H. L. Frisch	Time lag in transport theory. J. Chem. Phys. 36 (1962) 510.
71	H. L. Frisch	The time lag in diffusion. J. Phys. Chem. 61 (1957) 93.

72 G. R. Gavalas — Nonlinear differential equations of chemically reacting systems. (Heidelberg: Springer Verlag, 1968).

73 G. Gavalas and R. Aris — On the theory of reactions in continuous mixtures. Phil. Trans. Roy. Soc. A260 (1966) 351.

74 W. N. Gill, V. Ananthakrishnan and H. J. Barduhn, Laminar dispersion in capillaries. A.I.Ch.E. J. 11 (1965) 1063.

75 W. N. Gill and R. J. Nunge — Mechanisms affecting dispersion and miscible displacement. Ind. Eng. Chem. 61 (Pt. 9)(1969) 33.

76 H. J. Groenewold — The model in physics. In [68] p. 68.

77 L. Gross (ed.) — Symposium on sociological theory. (Evanston: Northwestern Univ. Press, 1959).

78 M. Gross — Mathematical models in linguistics. (Englewood Cliffs: Prentice-Hall, 1972).

79 G. G. Hall — Modelling--a philosophy for applied mathematicians. Bull. I.M.A. 8 (1972) 226.

80 J. M. Hammersley — Maxims for manipulators. Bull. I.M.A. 9 (1973) 276; 10 (1973) 368.

81 J. M. Hammersley — How is research done? Bull I.M.A. 9 (1973) 214.

82 J. M. Hammersley — Poking about for vital juices of mathematical research. Bull. I.M.A. 10 (1974) 235.

83 F. Harary — 'Cosi fan Tutte'--a structural study. Psych. Reports 13 (1963) 466.

84 F. Harary — Graph theory. (Reading: Addison-Wesley, 1969).

85 F. Harary, F. R. Norman and D. Cartwright, Structural models: an introduction to the theory of directed graphs. (New York: John Wiley, 1965).

86 R. Harré — An introduction to the logic of the sciences. (London, 1960).

87 R. Harré — The principles of scientific thinking. (London, 1970).

88 N. Hawkes (ed.) — International seminar on trends in mathematical modelling. (Lec. Notes in Econ. and Math. Systems 80). (Heidelberg: Springer Verlag, 1973).

89 M. B. Hesse — Operational definition and analogy in physical theories. Brit. Jnl. for the Phil. of Sci. 2 (1952) 281.

90 M. B. Hesse — Models in physics. Brit. Jnl. for the Phil. of Sci. 4 (1954) 198.

91 M. B. Hesse — Science and the human imagination: aspects of the history and logic of physical science. (London, 1954).

92 M. B. Hesse — Models and analogies in science. (London: Sheed and Ward, 1963).

93 M. B. Hesse — Models and analogy in science. In P. Edwards. (ed.)(1967) Vol. V.

94 M. Hesse — The structure of scientific inference. (Berkeley: Univ. of Cal. Press, 1974).

95 V. Hlaváĉek, M. Marek and M. Kubíĉek, Analysis of nonstationary heat and mass transfer in a porous catalyst particle. J. Cat. 15 (1969) 17, 31.

96 F. R. Hodson, D. G. Kendall and P. Tautu, Mathematics in the archaeological and historical sciences. (Edinburgh: Univ. Press, 1971).

97 E. H. Hutten — The role of models in physics. Brit. Jnl. for the Phil. of Sci. 4 (1954) 284.

98 E. H. Hutten — The ideas of physics. (London, 1967).

99 R. Issacs — Differential games. (New York: John Wiley, 1965).

100 K. Itô — On stochastic differential equations. Mem. Amer. Math. Soc. No. 4 (1951).

101 R. E. Kalman — On the mathematics of model building. In E. R. Caianello (ed.) Neural Networks. (Heidelberg: Springer Verlag, 1968).

102 A. Kaplan — The conduct of enquiry: methodology for behavioral science. (San Francisco: Chandler Pub. Co., 1964).

103 R. L. Kashyap and A. R. Rao Dynamic stochastic models from empirical data. (New York: Academic Press, 1976).

104 A. Kaufmann Introduction to the theory of fuzzy subsets. (New York: Academic Press, 1975).

105 H. B. Keller and D. S. Cohen Some positive problems suggested by nonlinear heat generation. J. Math. Mech. 16 (1967) 1361.

106 M. G. Kendall and A. Stuart The advanced theory of statistics. (New York: Hafner Pub. Co., 1963, 1966, 1967) 3 vols.

107 C. W. Kilmister and J. E. Reeve, Rational mechanics. (New York: American Elsevier, 1966).

108 A. Kuipers Model and insight. In [68] p. 125.

109 B. Lavenda, G. Nicolis and M. Herschkowitz-Kaufman, Chemical instabilities and relaxation oscillations. J. Theor. Biol. 32 (1971) 283.

110 P. F. Lazarsfeld (ed.) Mathematical thinking in the social sciences. (New York: Russell and Russell, 1969).

111 P. F. Lazarsfeld and N. W. Henry (eds.), Readings in mathematical social science. (Cambridge, Mass.: M.I.T. Press, 1966).

112 W. H. Leatherdale The role of analogy, model and metaphor in science. (Amsterdam and New York: North Holland/American Elsevier, 1974).

113 G. Levine and C. J. Burke Mathematical model techniques for learning theory. (New York: Academic Press, 1972).

114 R. Levins Evolution in changing environments. (Princeton: Princeton Univ. Press, 1968).

115 T-Y Li and J. A. Yorke Period three implied chaos. A.M.S. Monthly 82 (1975) 985.

116 M. J. Lighthill Keynote address to Int. Commission of Math. Instruction. Bull. I.M.A. 12 (1976) 87.

117 D. V. Lindley Introduction to probability and statistics. (Cambridge: Cambridge Univ. Press, 1965) 2 vols.

118 Y. A. Liu and L. Lapidus — Observer theory for lumping analysis of monomolecular reaction systems. A.I.Ch.E. J. 19 (1973) 467.

119 E. N. Lorentz — The problem of deducing the climate from the governing equations. Tellus 16 (1964) 1.

120 R. M. Loynes — The role of models. In [96] p. 547.

121 R. D. Luce and H. Raiffa — Games and decisions. (New York: John Wiley, 1957).

122 D. Luss — Some further observations concerning multiplicity and stability of distributed parameter systems. Chem. Eng. Sci. 26 (1971) 1713.

123 D. Luss and P. Hutchinson — Lumping of mixtures with many parallel first order reactions. Chem. Eng. J. 1 (1970) 129.

124 D. Luss and C. A. Pikios — Isothermal concentration oscillations on catalytic surfaces. Chem. Eng. Sci. 32 (1977) 191.

125 E. McMullin — What do models tell us?. In [186].

126 E. J. McShane — Stochastic calculus and stochastic models. (New York: Academic Press, 1974).

127 D. P. Maki and M. Thompson — Mathematical models and applications (with emphasis on the social, life and management sciences). (Englewood Cliffs: Prentice Hall, 1973).

128 J. E. Marsden and M. McCracken, The Hopf bifurcation and its applications (Appl. Math. Sci. 19). (Heidelberg: Springer Verlag, 1976),

129 R. M. May — Simple mathematical models with very complicated dynamics. Nature 261 (1976) 459.

130 R. M. May and G. F. Oster — Bifurcations and dynamic complexity in simple ecological models. Amer. Naturalist 110 (1976) 573.

131 D. H. Mellor — Models and analogies in science. Duhem vs. Campbell. Isis 59 (1968) 282.

132 H. D. Mellor — The matter of chance. (Cambridge: Cambridge Univ. Press, 1971).

133 E. Nagel, P. Suppes and A. Tarski (eds.), Logic, methodology and philosophy of science. (Stanford: Stanford Univ. Press, 1962).

134 C. V. Negoita and D. A. Ralescu, Applications of fuzzy sets to systems analysis. (New York: John Wiley, 1975).

135 G. Nicolis and J. Portnow Chemical oscillations. Chem. Rev. 73 (1973) 365.

136 B. Noble Applications of undergraduate mathematics in engineering. (New York: Macmillan, 1967).

137 G. C. Nooney Mathematical models, reality and results. J. Theoret. Biol. 9 (1965) 239.

138 U. Norlén Simulation model building. (New York: John Wiley, 1975).

139 M. F. Norman Markov processes and learning models. (New York: Academic Press, 1972).

140 O. Ore Graphs and their uses. (New York, Random House, 1963).

141 G. F. Oster, A. S. Perelson and A. Katchalsky, Network thermodynamics: dynamic modelling of biophysical systems. Quant. Rev. Biophys. 6 (1973) 1.

142 H. G. Othmer On the temporal characteristics of a model for the Zhabotinskii-Belousov reaction. Math. Biosci. 24 (1975) 205.

143 E. Parzen Modern probability theory and its applications. (New York: John Wiley, 1960).

144 R. Penrose The rôle of aesthetics in pure and applied mathematical research. Bull. I.M.A. 10 (1974) 266.

145 G. Polya How to solve it. (Princeton: Princeton Univ. Press, 1957). 2nd Edn. Doubleday Co., Inc., New York.

146 G. Polya Mathematics of plausible reasoning. I. Induction and analogy in mathematics; II. Patterns of plausible inference. (Princeton: Princeton Univ. Press, 1954) 2 vols.

147 G. Polya Mathematical discovery. (New York: John Wiley, 1962, 1965) 2 vols.

148 N. V. Prabhu — Queues and inventories. (New York: John Wiley, 1965).

149 A. Rapaport — N-person game theory: concepts and applications. (Ann Arbor: Univ. of Mich. Press, 1970).

150 A. Rapaport — Two person game theory: the essential ideas. (Ann Arbor: Univ. of Mich. Press, 1966).

151 W. H. Ray and J. R. Barney — The application of differential game theory to process control problems. Chem. Eng. J. 3 (1972) 237.

152 W. Regenass and R. Aris — Stability estimates for the stirred tank reactor. Chem. Eng. Sci. 20 (1965) 60.

153 W. T. Reid — Anatomy of the ordinary differential equation. Amer. Math. Monthly 82 (1975) 971.

154 M. Rosenblatt — Random processes. (Heidelberg: Springer Verlag, 1974).

155 A. Rosenblueth and N. Wiener — The roles of models in science. Phil. of Sci. 12 (1945) 316.

156 D. E. Rosner — The treatment of jump-conditions at phase boundaries and fluid dynamical discontinuities. Chem. Eng. Educ. 10 (1976) 190.

157 O. E. Rössler — Chaotic behavior in simple reaction systems. Z. Naturforsch. 31a (1976) 259.

158 M. F. Rubinstein — Patterns in problem solving. (Englewood Cliffs: Prentice Hall, 1975).

159 L. A. Segel — Simplification and scaling. SIAM Review 14 (1972) 547.

160 J. Serrin — Mathematical principles of classical fluid mechanics. Handbuch der Physik, Bd VIII/1. Eds. S. Flugge and C. Truesdell. (Berlin: Springer, 1959).

161 M. J. Sewell — Some mechanical examples of catastrophe theory. Bull. I.M.A. 12 (1976) 163.

162 J. A. Shohat and J. D. Tamarkin, The problem of moments. Amer. Math. Soc. (Math Surveys 1). (New York, 1943).

163 M. Sheintuch and R. A. Schmitz, Oscillations in catalytic reactions. Cat. Rev. 17 (1977) 119.

164 J. M. Smith Mathematical ideas in biology. (London: Cambridge Univ. Press, 1968).

165 J. M. Smith Models in ecology. (Cambridge: Cambridge Univ. Press, 1974).

166 T. T. Soong Randon differential equations in science and engineering. (New York: Academic Press, 1973).

167 M. Spector Models and theories. Brit. Jnl. for the Phil. of Sci. 16 (1965) 121.

168 D. J. Stewart (ed.) Automation theory and learning systems. (Washington, D.C.: Thompson Book Co., 1967).

169 P. Suppes A comparison of the meaning and uses of models in mathematics and the empirical sciences. In [68] p. 163.

170 J. Swanson On models. Brit. Jnl. for the Phil. of Sci. 17 (1966) 297.

171 R. D. Tanner Identification, hysteresis and discrimination in enzyme kinetic models. A.I.Ch.E. J. 18 (1972) 385.

172 A. Tarski A general method in proofs of undecidability. In [173].

173 A. Tarski, A. Mostowski and R. M. Robinson (eds.), Undecidable theories. (Amsterdam: North Holland Pub. Co., 1953).

174 G. I. Taylor Dispersion of a soluble matter in solvent flowing slowly through a tube. Proc. Roy. Soc. A219 (1953) 186.

175 G. I. Taylor Dispersion of salts injected into large pipes or the blood vessels of animals. Appl. Mech. Rev. 6 (1953) 265.

176 D. W. Theobald Models and methods. Phil. 39 (1964) 260.

177 R. Thom Structural stability and morphogenesis. (Reading: Benjamin, 1975).

178 J. R. R. Tolkien — Tree and leaf. (Boston: Houghton and Mifflin, 1965).

179 C. A. Truesdell and R. Toupin — The classical field theories. Handbuch der Physik, Bd. III/1. S. Flugge (ed.). (Berlin: Springer, 1960).

180 M. L. Tsetlin — Automaton theory and modelling of biological systems. (New York: Academic Press, 1973).

181 Ya. Z. Tsypkin — Foundations of the theory of learning systems. (New York: Academic Press, 1974).

182 J. J. Tyson — The Belousov-Zhabotinskii reaction. Lec. Notes in Biomath. 10. (Heidelberg: Springer Verlag, 1976).

183 A. Uppal, W. H. Ray and A. Poore, On the dynamic behavior of continuous stirred tank reactors. Chem. Eng. Sci. 29 (1974) 967.

184 A. Uppal, W. H. Ray and A. Poore, The classification of the dynamic behavior of continuous stirred tank reactors--influence of reactor residence time. Chem. Eng. Sci. 31 (1976) 205.

185 J. B. Ubbink — Model, description and knowledge. In [68] p. 178.

186 B. Van Rootselaar and J. F. Staal, Logic, methodology and philosophy of science. III. Proceedings of the third international congress for logic, methodology and philosophy of science. (Amsterdam, 1967).

187 J. Von Neumann and O. Morgenstern, Theory of games and economic behavior. (Princeton: Princeton Univ. Press, 1947).

188 J. Villadsen and W. E. Stewart, Graphical calculation of multiple steady states and effectiveness factors for porous catalysts. A.I.Ch.E. J. 15 (1969) 28.

189 J. Wei — Least squares fitting of an elephant. Chemtech. 5 (1975) 128.

190 J. Wei and J. C. W. Kuo — A lumping analysis in monomolecular reaction systems. Ind. Eng. Chem. Fundamentals 8 (1969) 114.

191 J. Williams — The compleat strategyst. (New York: McGraw Hill, 1954).

192 H. Wold — Bibliography on time series and stochastic processes. (Edinburgh: Oliver and Boyd, 1966).

193 E. Wong — Stochastic processes in information and dynamical systems. (New York: McGraw Hill, 1971).

194 J. Worster — A discovery in analysis aided by a computer. Bull. I.M.A. 9 (1973) 320.

195 L. A. Zadeh — Fuzzy sets. Information and Control 8 (1965) 338.

196 L. A. Zadeh, K-S Fu, K. Tanaka and M. Shimura (eds.), Fuzzy sets and their applications to cognitive and decision processes. (New York: Academic Press, 1975).

197 E. C. Zeeman — Catastrophe theory. Sci. Amer. 234, No. 4 (1976) 65.

198 J. A. Andrews and R. R. McLone, Mathematical modelling. (London: Butterworths, 1976).

199 X. J. R. Avula (ed.) — Proceeding of the first international conference on mathematical modelling. Univ. of Missouri, Rolla (1977) 5 vols.

200 H. T. Banks — Modelling and control in the biomedical sciences. (New York: Springer Verlag, 1975).

201 H. A. Becker — Dimensionless parameters: theory and methodology. (New York: John Wiley, 1976).

202 D. F. Boucher and G. E. Alves — Dimensionless numbers. Chem. Eng. Prog. 55 (1959) 55.

203 P. W. Bridgeman — Dimensional analysis. (New Haven: Yale Univ. Press, 1922).

204 T. Bröcker — Differentiable germs and catastrophes. London Math. Soc. Lec. Note Ser. 17. (Cambridge: Cambridge Univ. Press, 1975).

205 D. R. J. Chillingworth — Differential topology with a view to applications. (London: Pitman, 1976).

206 D. S. Cohen and A. Poore — Tubular chemical reactors: the "lumping approximation" and bifurcation of oscillatroy states. SIAM J. on App. Math. 27 (1974) 416.

207 W. J. Duncan — Physical similarity and dimensional analysis. (London: Edward Arnold, 1953).

208 R. Esnault-Pelterie — L'analyse dimensionelle. (Paris: Ed. F. Range, 1945).

209 C. M. Focken — Dimensional analyses and their applications. (London: Edward Arnold, 1953).

210 W. N. Gill, E. Ruckenstein and H. P. Hsieh, Homogeneous models for porous catalysts and tubular reactors with heterogeneous reactions. Chem. Eng. Sci. 30 (1975) 685.

211 J. Guckenheimer, G. Oster and A. Ipaktchi, The dynamics of density dependent population models. J. Math. Biol. 4 (1977) 101.

212 A. A. Gukhman — Introduction to the theory of similarity. (New York: Academic Press, 1965).

213 R. Haberman — Mathematical models. (Englewood Cliffs: Prentice-Hall, 1977).

214 J. L. Howland and C. A. Grobe — A mathematical approach to biology. (Lexington, Mass.: Heath, 1972).

215 H. E. Huntley — Dimensional analysis. (Rinehart, 1951).

216 D. C. Ipsen — Units, dimensions and dimensionless numbers. (New York: McGraw-Hill, 1960).

217 E. W. Jupp — An introduction to dimensional method.

218 S. J. Kline — Similitude and approximation theory. (New York: McGraw-Hill, 1965).

219 N. Koppel and L. N. Howard — Plane wave solutions to reaction-diffusion equations. Stud. Appl. Math. 52 (1973) 291.

220 F. W. Lanchester — Theory of dimensions and its applications for engineers. (London: Crosby, Lockwood, 1936).

221 H. Langhaar — Dimensional analysis and the theory of models. (New York: Wiley, 1951).

222 T-Y Li and J. A. Yorke — The "simplest" dynamical system. In Dynamical Systems, Vol. 2. (New York: Academic Press, 1976).

223 C. C. Lin and L. A. Segel — Mathematics applied to deterministic problems in the natural sciences. (New York: Macmillan, 1974).

224 K. Z. Lorenz — Fashionable fallacy of dispensing with description. Naturwiss. 36 (1973) 1.

225 Y-C Lu — Singularity theory and an introduction to catastrophe theory. (New York: Springer Verlag, 1976).

226 V. L. Makarov and A. M. Rubinov, Mathematical thoery of economic dynamics and equilibria. (New York: Springer Verlag, 1977).

227 T. Maruyama — Stochastic problems in population genetics. (New York: Springer Verlag, 1977).

228 C. A. Mihram — A closer look at Thom's catastrophe theory. Proc. 1st Int. Conf. on Math. Modelling. X. J. R. Avula (ed.) Vol. 1 (1977) 359.

229 G. Murphy — Similitude in engineering. (New York: Ronald Press, 1950).

230 J. Palacios — Dimensional analysis. (New York: Macmillan, 1964).

231 R. C. Pankhurst — Dimensional analysis and scale factors.

232 J. Pawlowski — Die Ähnlichkeitstheorie in der physikalisch-technischen Forchung. (Heidelberg: Springer Verlag, 1971).

233 A. W. Porter — The method of dimensions. (London: Methuen, 1946).

234 T. Poston and I. N. Stewart — Taylor expansions and catastrophes. (London: Pitman, 1976).

235 T. Poston and I. N. Stewart — Catastrophe theory and its applications: an advanced outline survey. (London: Pitman, 1977).

236 L. Rayleigh — Scientific papers. (Cambridge: Cambridge Univ. Press, 1899-1919). 1, 377; 4, 452; 6, 300.

237 M. Reiner — The Deborah number. Physics Today Jan. '64 (1964) 62.

238 H. Rouse — Compendium of Meterology. T. F. Malone (ed.). (Boston: Amer. Met. Soc., 1951).

239 L. I. Sedov — Similarity and dimensional methods in mechanics. (New York: Academic Press, 1959).

240 L. A. Segel — Mathematics applied to continuum mechanics. (New York: Macmillan, 1977).

241 R. Shinnar and P. Naor — Residence time distribution in systems with internal reflux. Chem. Eng. Sci. 22 (1967) 1369.

242 R. Shinnar, F. J. Krambeck and S. Katz, Interpretation of tracer experiments in systems with fluctuating throughput. Ind. Eng. Chem. Fundamentals 8 (1969) 431.

243 R. Shinnar, P. Naor and S. Katz, Interpretation and evaluation of multiple tracer experiments. Chem. Eng. Sci. 27 (1972) 1627.

244 R. Shinnar, D. Glasser and S. Katz, The measurement and interpretation of contact time distributions for catalytic reactor characterization. Ind. Eng. Chem. Fundamentals 12 (1973) 165.

245 R. Shinnar and Y. Zvirin — A comparison of lumped-parameter and diffusional models describing the effects of outlet boundary conditions on the mixing in flow systems. Water Res. 10 (1976) 765.

246 R. Shinnar and Y. Zvirin — Interpretation of internal tracer experiments and local sojourn time distributions. Int. J. of Multiphase Flow 2 (1976) 495.

247 V. E. J. Skoglund — Similitude: theory and applications. (New York: Int. Textbooks, 1967).

248 D. L. Solomon and C. Walter (eds.), Mathematical models in biological discovery. (New York: Springer Verlag, 1977).

249 S. Smale — A mathematical model of two cells via Turing's equation. Lec. on Math. in Life Sciences. Amer. Math. Soc. 6 (1971) 17.

250 R. Thom Structural stability, catastrophe theory and applied mathematics. SIAM Rev. 19 (1977) 189.

251 J. M. T. Thompson Catastrophe theory and its role in applied mechanics. Theor. and App. Mech. (Proc. 14th IUTAM Congress Delft. Ed. W. T. Koiter) (1976) 451.

252 J. M. T. Thompson and G. W. Hunt, The instability of evolving systems. Interdisciplinary Sci. Revs. 2 (1977) 240.

253 C. Van Heerden Autothermic processes. Ind. Eng. Chem. 45 (1953) 1242.

254 A. T. Winfree Spiral waves of chemical activity. Science 175 (1972) 634.

255 A. E. R. Woodcock and T. Poston, A geometrical study of the elementary catastrophes. (New York: Springer Verlag, 1974).

256 L. C. Woods The art of physical and mathematical modelling. In A spectrum of mathematics. K. Butcher (ed.) 142. (Auckland: Auckland Univ. Press, 1971).

257 J. M. Ziman Mathematical models and physical toys. Nature 206 (1965) 1187.

258 N. R. Amundson and O. Bilous Chemical reactor stability and sensti sensitivity. A.I.Ch.E. J. 1 (1955) 513.

259 G. Birkhoff Hydrodynamics: a study in logic, fact and similitude. (Princeton: Princeton Univ. Press, 1960).

260 N. Campbell Dimensional analysis. Phil. Mag. 47 (1924) 481.

261 J. Villadsen and M. L. Michelsen, Solution of differential equation models by polynomial approximation. (Englewood Cliffs: Prentice Hall, 1978).

Subject index

Name index

Re, k and π : a conversation on some aspects of mathematical modelling

Rutherford Aris

Department of Chemical Engineering and Materials Science, University of Minnesota, Minneapolis, Minnesota 55455, USA
(Received 20 June 1977)

Re and k discuss π 's approach to what we understand by a mathematical model; how it may have certain generic properties and how hierarchies of related models arise in connexion with a given physical situation. Joined at this point by π, they continue to talk about the way in which models are formulated and prepared for solution. This preparation involves such things as the choice of the most suitable dimensionless variables, reduction to the smallest number of equations, proving uniqueness and discovering the shape of the solution. In conclusion, some aspects of the presentation of the results are discussed. The abbreviated names of Reynolds, Boltzmann and Pythagoras have been used only to denote the engineer, natural scientist and mathematician taking part in the discussion; there is no allusion to the points of view of the historic figures.

There is a campus not so far from a distant ocean which breathes, as far as that is possible in this world of sin and smog, the quiet air of active leisure that bequeathed its name to scholarship. It is young enough by the standards of the Old World though old in relation to the civilized history of the New. Time will soon forgive its slight architectural promiscuities and, if the climate permit, a century or two of mowing, rolling and watering will turn the lawns from baize to velvet. But already its trees are venerable, respected by the architect and well cared for when the weight of years threatens their integrity. Nor is the ambience of this institution its only claim to distinction. What it does it does exceeding wèll and holds high the torch that it passes on so that all who are, or have been, related to it murmur their *et ego's* with mingled pride and humility.

There it was, on an 'unusually' warm day in the late summer, that I overheard the conversation of three friends. One appeared to be an engineer whom I will call Re, a useful enough fellow, positive but of no particular magnitude. His friend, *k*, was more of a physics or chemistry type, a brilliant but empirical chap who was reputed to have done little enough work for his degree*. They were joined, as you shall hear, by π, a rather portly man, distinctly older than the others, very precise in his determination, but not quite rational for all that—in short a typical mathematician.

'Here's some shade' said *k* as they came up a short slope to the base of a fine old oak whose branches spread over the adjacent grass.

'Your plane tree, I suppose,' said Re, 'but with no *agnus castus*[1] to guard your chastity, nor chorus of cicadae for that matter'.

'Don't be a jackass.' His friend's accent and usage indicated that he had been brought up in England and he seemed somewhat unreceptive to Re's conceits. 'This is an Engleman oak and anyway there's no stream to cool your feet. Forget Achelous and the Nymphs and tell me what π's been saying about mathematical modelling for I see you have your notebook in your left hand.'

'Well, he makes some pretty strong claim for models. They have a life of their own, sez'ee; not just a question of being derived from the physical situation.'

'That sounds like him alright, but I bet that came at the end of the argument—"Thus we see that, like ... models have a life of their own". How did he begin?'

'With the O.E.D. of course. That man has more faith in the dictionary than Isidore himself. "Model" it appears comes from "modus", a measure, through Italian and French with very little change into all the Germanic languages. It is basically a representation of

*1.36×10^{-16} ergs to be precise.

structure, as in an architect's plan or mock up. By transference it can be used for a summary or outline of a literary work. In general, it is a more or less accurate representation on a different scale. A second class of meanings centre round the idea of a type of design as it was used in Cromwell's New Model, his reorganization of the Parliamentary army. The third class has to do with objects of imitation or exemplars, as in "the very model of a modern major-general". Then there are one or two peculiar meanings that old π couldn't resist wasting five minutes on.'

'Typical!' said k with emphasis. 'I suppose mathematical model fits into the first category.'

'Precisely,' Re replied, 'the idea being that the change of scale is not a physical one, but rather a difference in the level of abstraction. A mathematical model generally represents its prototype in a set of equations and is good in so far as it has caught the essence and ignored the accidents of the original.'

'But model theory in logic stands that on its head. There a model is, if anything, less abstract and more concrete than the axioms which are fulfilled in it.'

'Yes, he made the point that there was a certain reciprocity about the notion of a model. An aeroplane is a model of the equations which have to be solved in designing it just as much as the equations are a model of the physical entity. 'Like Menander and life', old π said with an extra puff around the gills, 'it is at times difficult to know which is imitating the other'."*

'He is a bit much at times, isn't he. But I suppose case is obvious enough for a mathematical model; no one's going to confuse the hawk of a differential equation with the handsaw of a chemical reactor.'

'No, but more to the point was a distinction that Maynard Smith has made between models and simulations[3]. Smith uses the term "simulation" for a mathematical description adapted to some practical end. The value of such a description increases as more and more detail is incorporated and its agreement with a particular situation becomes more and more exact. By contrast, a mathematical description with a theoretical purpose is valuable when it incorporates as little specific detail as possible and gives results that are broadly true for whole classes of situations. For example, chaotic solutions obtain for a whole class of population models of the type $N_{t+1} = f(N_t)$, where the population in year $t + 1$ is a nonlinear function of the population in year t. When the function f is...'

'Hang on, what's a chaotic solution?' interrupted k.

'Haven't you read May's brilliant review article in Nature,[1,4] said his friend, 'or Li and Yorke on period 3 implies chaos[5]? Chaos is the name that has been given to the situation when there are periodic solutions of all periods as well as solutions which are not periodic nor asymptotic to any periodic solution even though there may be stable periodic solutions around.'

'I don't quite get that.'

'Well, suppose x_n is the population at the nth point in time; that is, time is divided into stages of duration Δt—years, breeding seasons or whatever the appropriate interval may be—and one value of the dependent variable (for example, population) characterises the whole interval from $n\Delta t$ to $(n + 1)\Delta t$. Call it x_n. Then x_n is governed by some equation $x_{n+1} = f(x_n)$ which allows x_1 to be calculated from x_0, x_2 from x_1 and so on. If there is a particular starting point x_0 such that calculating $x_1 = f(x_0)$, $x_2 = f(x_1), \ldots$ leads back after p steps to x_0 (i.e. $x_p = x_0$), then any of the points $x_0, x_1, \ldots x_{p-1}$ is said to be periodic with period p, and the set of points $\{x_0, x_1, \ldots x_{p-1}\}$ is the orbit of the solution. A solution of period 1 is a fixed point for $x_0 = x_1 = x_2 = \ldots$; a solution of period 2 has two values and jumps forth and back between them $x_0 = x_2 = x_4 = \ldots$ while $x_1 = x_3 = x_5 = \ldots$. If a periodic solution is stable then, for a starting point y_0 sufficiently close to x_0, the sequence $y_1 = f(y_0)$, $y_2 = f(y_1), \ldots$ will converge to the periodic solution; that is $|y_n - x_n|$ approaches zero as n tends to infinity or both lim inf and lim sup of $|y_n - x_n|$ are zero. Now a chaotic solution may start very close to a periodic solution and as n increases it may often be near the periodic solution, that is $|y_n - x_n|$ may have a lower limit of zero, $\liminf |y_n - x_n| = 0$, but the upper limit, $\limsup |y_n - z_n|$, is definitely not zero. This means that the chaotic solution may so to speak "hang around" with the periodic one and spend long periods of time in its company but ever and anon it will go off by itself in a totally unrelated way. Li and Yorke have proved that whenever there is a solution of period 3 there are also solutions of all periods and a swarm of chaotic solutions as well.'

'Very interesting. But that sounds like one of π's famous digressions where he rumbles through the undergrowth and comes out with his hair full of leaves wondering why he got sidetracked.'

'Well it was in a way, but he used it to illustrate the idea of generic properties that models can discover. This sort of thing happens for any $f(x)$ with a tuneable hump.'

'What the dickens is a "tuneable hump"?'

'Oh, a function like $f(x) = \lambda x(1 - x)$ which has a maximum in the interval $(0, 1)$ at $x = 1/2$ and a parameter like λ that can be 'tuned' (cf. *Figure 1*). For $\lambda \prec 1$ the equation $x_{n+1} = \lambda x_n(1 - x_n)$ only has the trivial steady state solution $x = 0$ and all solutions converge to it. For $1 \prec \lambda \prec 3$ the unique, non-trivial stable solution moves away from zero, but it becomes unstable for $\lambda \succ 3$ and two period 2 solutions arise and become stable. However, their stability breaks

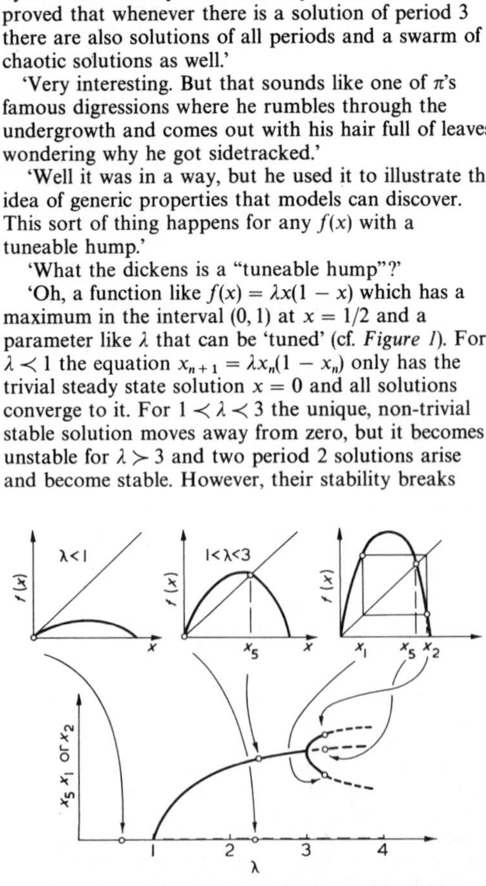

Figure 1 Steady state solutions and solutions of period 2 with $f(x) = \lambda x(1 - x)$

*'Ω *Μένανδρε κᾱι βίε, πότερος 'άρ' 'νμῶν πότερον 'απεμιμήσατο.*
Aristophanes the Grammarian

down at $\lambda = 3.45$ and they bifurcate into four of period 4; their stability doesn't last long however and...'

'O.K. enough is enough'.

'I wasn't even halfway through; but anyhow, you can read it for yourself in May's review or in May and Oster's paper[6].'

'Get back on track and tell me where this is all leading' said k after noting down the references.

Re continued 'After the point about generic properties π pointed out that models don't arise in isolation, both the conditions of modelling and the purpose for which it is being done determine the appropriateness of the model—its Watkins number, so to speak*. For example, in trying to understand convective diffusion there are a number of models of different degrees of sophistication. The basic problem can be illustrated by thinking of laminar flow through a long cylindrical tube. The velocity varies from a high of twice the mean speed u on the centre line to zero at the wall—in fact it is given by $2u(1 - r^2/a^2)$ where r is the radial distance from the centre line and a the tube's radius. If there were no diffusion a thin disk of dye would be drawn out into a continually lengthening paraboloidal shell or a plane front with dyed fluid coming from behind would be pushed out into a parabolic nose. But, in the presence of molecular diffusion, such a paraboloid could not retain a sharp edge. Molecules of dye would diffuse outwards from the fast moving tip to the slower adjacent streams and water molecules near the wall diffuse toward the faster central streams, diluting the dye. So, very little of the dye moves with anything like the extreme speeds of zero and $2u$. In fact, the centre of gravity of the dye tends to move with the mean speed of the stream and the mean concentration of the dye across the tube spreads out about the centre of gravity just like the concentration would were the velocity uniform (i.e. plug flow at velocity, u) and there were an effective longitudinal dispersion coefficient D_e. This is an empirical observation and was made first by a physiologist who claimed it was obvious. Sir Geoffrey Taylor[7] in his original analysis remarks that the only difficult thing to understand is the sentence following the words "it is obvious that..."

'That's the disease of mathematicians rather than physiologists, I should have thought', interjected k.

'I know', said Re, 'but it's catching. Anyway Taylor showed that it was obvious though it took half the paper to prove it and show that D_e was inversely proportional to the molecular diffusion coefficient D.

'Now suppose the physical system is D: the most complete mathematical model that can be derived for it is Δ_1, a parabolic partial differential equation for the concentration of the dye as a function of x, the distance along the axis from the original plane interface, r, the radial distance from the axis and the time, t. This is a complete model in that it involves the fewest and most acceptable hypotheses: the basic principles of physics such as the conservation of matter, simple constitutive relations such as Fick's law, boundary conditions such as the impermeability of the tube wall. This full model may be moulded to some extent, say by taking an origin moving with the mean speed of the flow $y = x - ut(\Delta_1')$ or by a change of variables adapted to the boundary and initial conditions (Δ_2), but it presents essentially the same aspect and demands a rather complete numerical solution before yielding results that can be compared with the phenomena. The purpose for which the model is intended now becomes a factor. If very precise values of the concentration are required it may be necessary to use such a model and put up with the ımerical complexities that it entails. But if the purpose is less exacting, say, to get a good idea of how the dispersive process affects the mean concentration across the tube it may not be necessary to solve the full equations. Instead of an equation for the concentration at each point we content ourselves with a system of equations for the moments:

$$\int_{-\infty}^{\infty} y^p c(y, r, t)\,dy = c^{(p)} \quad \text{or} \quad m^{(p)}(t) = \int_0^a 2rc^{(p)}\,dr/a^2$$

The equations for these should be simpler since they involve fewer variables and it turns out that they can be solved in the sequence $m^{(0)}$, $c^{(0)}$, $m^{(1)}$, $c^{(1)}$, $m^{(2)}$. Now $m^{(0)}$ is just the total amount of solute present and it is not surprising that integrating the equations gives $dm^{(0)}/dt = 0$, that is, the total amount doesn't change. The equation for $c^{(0)}$ is more difficult and its solution involves Fourier–Bessel series. The solution for $m^{(1)}$ shows it to be asymptotically constant which means that the centre of gravity of the solute eventually moves with the mean speed of the flow. The constant that $m^{(1)}$ approaches depends on the initial distribution of the solute and the solution for $c^{(1)}$ again requires a series of Bessel functions. But the equation for $m^{(2)}$ shows that $m^{(2)}$ tends to become linear in t. Now by comparison with the plug flow case $m^{(2)} = 2D_e t$ and it turns out that $D_e = D + a^2u^2/48D$. This model is Δ_3 and it must be emphasised that its results are approximate in two ways. They are in principle approximate in that they can only provide a finite number of moments of the distribution of solute and they are in practice approximate since the computation of Fourier–Bessel series is cumbersome and the simple results are asymptotic statements: the centre of gravity approaches the mean speed and the rate of growth of half the variance approaches the dispersion coefficient D_e'[8].

'That's quite a mouthful and much enhanced by the arm waving' said k.

'Oh, I haven't finished yet. Let's call the model with plug flow and a dispersion coefficient Δ_4 (that is, Δ_3 shorn of its refinements). Also we note that if $P = au/D \gg 1$ (say > 100) then $D_e = a^2u^2/48D$ since the D in $D_e = D + a^2u^2/48D$ is negligible. This is Taylor's original result obtained by a different way of approximating Δ_1. If P has an intermediate value then both terms are needed. If $P = au/D \ll 1$ (say < 0.2) then the term D in the expression for D_e dominates. Call these cases Δ_4', Δ_4'', and Δ_4''' respectively. In addition, let Δ_5 be the case of pure convection with no term at all for diffusion, and Δ_6 an empirical formula found by Gill *et al.*[9] to be an excellent approximation to the exact solution of Δ_1. Then we can arrange these models in relation to the conditions and purpose (as in *Figure 2*). First, we agree on a sufficient level of

* 'He was always appropriate and it was fitting that none should see him dead'—concluding lines 'Gino Watkins'; biography of J. M. Scotts'

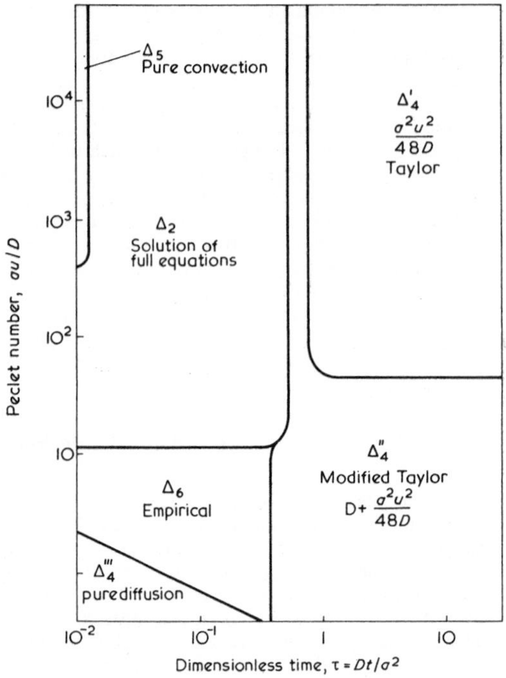

Figure 2 Domains of applicability of different models of Taylor diffusion

accuracy; this is arbitrary and some may be more exacting than others in their demands but for the moment this level is fixed. The conditions may be represented by $P = au/D$, the so-called Peclet number. The purpose may be indicated by the dimensionless time $\tau = Dt/a^2$; that is, we intend to account for the behaviour of the system at short or at long times. The P, τ-plane can then be divided into regions in each of which a given model is adequate. For example, with high P and low τ there is little diffusion and no time for it to work anyway, so Δ_5, the pure convection model is good. The Δ_4 models are by nature asymptotic so we expect to find their regions where τ is large, and for reasons just given Δ_4' will be above Δ_4'' (i.e. Δ_4' is where P is larger). The detailed equations Δ_1 may be needed for smaller τ and larger P and the interpolatory model Δ_6. The boundaries of all these regions are fuzzy or rather the regions overlap for two models could give the same accuracy at certain points. Indeed, Δ_2 presumably applies everywhere since it claims to be full and true; but even Δ_2 would break down for sufficiently short times since Fick's law would have to be replaced by some molecular description. If we set up more exacting standards of accuracy the domains of Δ_4, Δ_5 and Δ_6 would move outwards in the $\ln P, \ln \tau$-plane leaving more and more of the plane to Δ_2. If we let accuracy go to pot the regions would overlap more and more since it would not matter which model were used. In fact, if we made stereographic projections of the regions on a sphere whose radius was a measure of the accuracy demanded, we could get a very neat three-dimensional picture with the Δ_4, Δ_5, Δ_6 regions contracting on the polar axis as the exactingness increased.'

'Hold hard a minute', said k, raising a hand in protest, 'you'll be wanting coloured cine-holography next.'

'That would be rather fun, wouldn't it,' Re replied. 'But seriously, π claims that models must be viewed in this kind of relationship and not in isolation. Later models (Gill and Shankar[10] have one with a time dependent dispersion coefficient) can then be fitted in to their place.'

'Yes, that's obvious enough. But what about the origin of models and their intrinsic relations. The "exact" model Δ_1 comes from the physical system D. Δ_3, the set of equations for moments, is obtained from it by the mathematical operation of multiplying by y^p and integrating or averaging. To apply it a physical observation, or analogy, is made; one that calls attention to the fact that plug flow with longitudinal dispersion produces the same sort of effect. This is Δ_4 and Δ_4'' is its expression with Δ_4' and Δ_4''' as limiting cases. Suppose some totally different analogy were made, say with a sequence of stirred tanks with flow from one to the next. A sharp burst of tracer would pass through such a system with some mean speed and the variance of the tracer would increase linearly but a totally different set of equations would be involved. How would such a model (call it T in contrast to the Δ's) be related to Δ_4?'

'I suppose π would say that T and Δ_4 were cognate, being born of the same physical system D, but not immediately derivable one from the other. Actually the stirred tank sequence would be hard to justify for the open tube with laminar flow but would be easy to motivate for a packed bed.'

'You speak like an engineer,' said k, 'How so?'

'Well, if you have a cylinder packed with small spheres they pack down and form small chambers between successive layers. As the fluid pushes through the bed it sort of jets into these cavities from the previous layer, stirring them up. This makes the physical analogy of a chain of stirred tanks a very natural one.'

'Fair enough, we have hierarchies of models cognate to or derived from one another and related to a physical system as prototype. What does π have to say about types of models?'

'Oh nothing original; just the usual types of mathematical equation—ordinary or partial differential equations, difference equations, *etcetera, etcetera, etcetera.* He did make a point about verbal models and mechanical analogies that Duhem derided so hugely[11]. Old π had great fun quoting him at length,

> "The employment of similar mechanical models... is a regular feature of the English treatises on physics. Here is a book intended to expound the modern theories of electricity... in it there are nothing but strings which move around pulleys, which roll around drums, which go through pearl beads, which carry weights; and tubes which pump water while others swell and contract; toothed wheels which are geared to one another and engage hooks. We thought we were entering the tranquil and neatly ordered abode of reason, but we find ourselves in a factory."

'Yes he'd enjoy that,' said k.

'I did. I did.' It was a new voice, for whilst they were busy talking π had come up and was resting his portly frame on the tree trunk behind them. No end chuffed he was to think that anyone would discuss his lectures outside of class and he lowered his bulk to the ground prepared to take over the conversation. 'Great fellow Duhem full of ire and irony, not to mention the rather extravagant eloquence of the French. Touch of bombast of course. Know what "bombast" comes from?'

'Cotton wool as padding, of course,' said Re hastily fearing a long digression on words derived from names of textiles. 'Not unrelated to "fustian", but not as some people think derived from the full name of Paracelsus,' he continued, 'but get back to mathematical models'.

'I was only making the point that language itself carries about little models in its words, using them metaphorically so that they sparkle with inner meanings—"a kist o ferlies" as Douglas Young called his dictionary of the older Scottish tongue*.'

'But what, sir, do you see as the next important aspect of the craft of modelling?' k made a last ditch attempt to get π back on track.

'The formulation of models of course,' said π. 'We have to distinguish between physical laws, such as the conservation of mass, energy or momentum, and constitutive relations, such as Fick's law or the stress–strain relations of a solid. The former are generally applicable to all situations or at any rate to a whole class, such as non-relativistic situations; the latter are dictated by specific assumptions about particular species. If F is the net flux of an entity into a given region and G the rate of generation there while H is the total amount of a given entity there, the conservation principle is:

$$F + G = \mathrm{d}H/\mathrm{d}t$$

If the system is distributed $\boldsymbol{f}$ is the flux vector per unit area and g is the rate of generation per unit volume while h is the concentration of the entity, then we can take an arbitrary region Ω with surface $\partial\Omega$ and $F = -\iint \boldsymbol{f}.\boldsymbol{n}\,\mathrm{d}S$, where $\boldsymbol{n}$ is the outward normal to $\partial\Omega$, $G = \iiint g\,\mathrm{d}v$ and $H = \iiint h\,\mathrm{d}V$. Then using Green's theorem and assuming sufficient continuity gives:

$$-\nabla.\boldsymbol{f} + g = \frac{\partial h}{\partial t}$$

The constitutive relation comes in to relate $\boldsymbol{f}$, g and h to each other. For example in a diffusion situation g might be zero, h the molar concentration c and $\boldsymbol{f}$ given by Fick's law as $-D\nabla c$. Then the combination of conservation principle and constitutive relation would give an equation for c:

$$\frac{\partial c}{\partial t} = \nabla.(D\nabla c)$$

In another situation there might be adsorption onto a surface and $h = c + \alpha n(c)$, where α is the surface area per unit volume and $n(c)$ the equilibrium surface concentration. If there is surface, as well as bulk, diffusion $\boldsymbol{f}$ might be $-D\nabla c - \alpha D_s\nabla n$ giving;

$$[1 + \alpha n'(c)]\frac{\partial c}{\partial t} = \nabla\cdot[D + \alpha D_s n'(c)]\nabla c$$

a very different equation on account of the very different constitutive relation.'

'That is clear enough,' said Re, 'but where do you go from there?'

'You will see if you keep coming to my lectures,' π rejoined, 'but to put you out of your agony of suspense let me outline things for you.'

'I wish you would be good enough to give us a really detailed account—unless you are pressed for time,' said k rolling into a more comfortable position on the grass.

But his slight sarcasm and the minor impertinence of the allusion[12] were both lost on π who continued loftily. 'I cannot, of course, do justice to the course in the short time I have available, but one should at this point spend some time on the formulation of boundary conditions and say something about lumped and distributed models. The bulk of the course will then be on the just moulding of models and their presentation. The latter calls for all the graphic imagination one can muster for there is nothing like a picture or diagram for getting a result across. Mathematicians and physicists have something to learn from engineers there.'

'There!?!' said Re to himself with an internally raised eyebrow but aloud his question was, 'What do you mean by the moulding of models?'

'I thought you'd never ask,' beamed π ignoring the fact that he'd given the others little enough chance. 'That's the core of my course for I thought it would be good to bring together a lot of the methods that one uses only half consciously. An obvious task is to simplify the system of equations and to scale them appropriately. Everyone does this with whatever grace and style they can command but the principles had rarely been explicated before Segel's SIAM Review paper on scaling[13] and his book with Lin on "Mathematics applied to deterministic problems in the natural sciences".[14] The general principle is that the burden of showing the relative magnitude of the important quantities should be borne by the dimensionless parameters and that the dimensionless variables should be scaled to the interval (0, 1) or thereabouts.'

But why bother to make the variables dimensionless?' objected k. Surely it is better to keep as much of the physical magnitude and meaning as possible.'

'You must not confuse meaning and magnitude. The dimensionless variables retain their physical meaning. In fact, they gain in meaning, for they are being compared with some quantity of the same dimensions which are characteristic of the system. A quantity with dimensions is also being compared with some other quantity of the same dimensions but the second quantity is arbitrary—some sacred lump of platinum laid up somewhere. Metricization may be good for merchants and mechanics; intellectually it is neither here nor there and the effort would be far better spent in restoring Latin as the lingua franca of the educated.

* "I'm canty yet wi sma delytes, albeid ma baird's sae black and swack. I ken a thing that's like a kist of ferlies gif ye read." From D. C. C. Young's poem "Thesaurus Paleo-Scoticus"

When you make a number or variable dimensionless you invest it with a meaning intrinsic to the problem—also you get the smallest number of parameters.'

'Can't you get that from the Buckingham π theorem?' Re asked.

'Yes, but that's all you can get. You can only start moulding the equations when you have the equations. To set down the variables and quantities will tell you how many independent dimensionless groups may arise but not where they should be put. There are trade-offs between the equations and their boundary conditions. For example[15] if you have the problem of a reactant diffusing into a slab of porous catalyst and there disappearing by first order reaction which has the equations:

$$D\frac{d^2c}{dx^2} = kc \quad \text{in} \quad |x| < L,$$

$$c = c_0 \quad \text{on} \quad |x| = L;$$

you can take $u = c/c_0$ and $y = x/L$ to give:

$$\frac{d^2u}{dy^2} = \left(\frac{kL^2}{D}\right)u \text{ in } |y| < 1$$

$$u = 1 \qquad \text{on} \quad |y| = 1$$

and there is a single dimensionless group $\phi^2 = kL^2/D$ in the equation. If $(D/k)^{1/2}$ is used as the characteristic length and $z = x(k/D)^{1/2}$ then

$$\frac{d^2u}{dz^2} = u \quad \text{in} \quad |z| < \phi$$

$$u = 1 \quad \text{on} \quad |z| = \phi$$

Again there is only a single parameter but this time it is in the boundary condition, in fact, in the position of the boundary.'

'Fair enough,' said k. 'So you get the variables and parameters into dimensionless form and have the smallest number of each, how do you know you have the smallest number of equations?'

'There again these are trade-offs,' said π. 'Take the non-isothermal case of the same problem where k is a function of the temperature T and in addition to:

$$D\frac{d^2c}{dx^2} = k(T)c$$

we have:

$$\lambda\frac{d^2T}{dx^2} = (\Delta H)k(T)c$$

as a differential equation for T; where (ΔH) is a heat of reaction and λ a conductivity. If λ, ΔH and D are constant then the combination $D(-\Delta H)c + \lambda T = W$ (say) satisfies the simple equation:

$$\frac{d^2W}{dx^2} = 0$$

This means W is a linear function of x and with the right boundary conditions may be a constant. For example, if we seek symmetrical solutions or solutions with the same values of c and T—and hence of W—at $x = \pm L$, then $D(-\Delta H)c + \lambda T = D(-\Delta H)c_0 + \lambda T_0$, a constant. This fact is recognized if we put:

$$c = c_0(1 - v), \qquad T = T_0(1 + \beta v),$$

$$\beta = D(-\Delta H)c_0/\lambda T_0$$

But substituting these into either equation now gives the same equation for v, namely:

$$\frac{d^2v}{dx^2} + \frac{1}{D}k(T_0(1 + \beta v))(1 - v) = 0, \qquad x < L$$

$$v = 0, \quad |x| = L$$

If we make x dimensionless by $x = Ly$ and let:

$$\phi^2 = k(T_0)L^2/D$$

$$f(v) = [k(T_0(1 + \beta v))/k(T_0)](1 - v)$$

then:

$$\frac{d^2v}{dy^2} + \phi^2 f(v) = 0, \quad |y| < 1$$

$$v = 0, \quad |y| = 1$$

[*Figure 3* shows the possible forms of $f(v)$]

'Let me interrupt', said Re, 'and just say that you have made your point adequately enough. You've got the smallest numbers of equations with the smallest number of parameters and let's suppose there's no simple analytical solution. Then I presume you have to go off to the computer and compute the real answer.'

'Good heavens, NO!' π almost shouted. 'Don't let the genius of this place even hear you think such a thing. Rushing off to computers is a criminal act and almost justifies the cynics who say that all computers should be built with an on-line incinerator. No, no, no, it's only safe to compute when you know the answer, or, at least have found out as much as possible about it.'

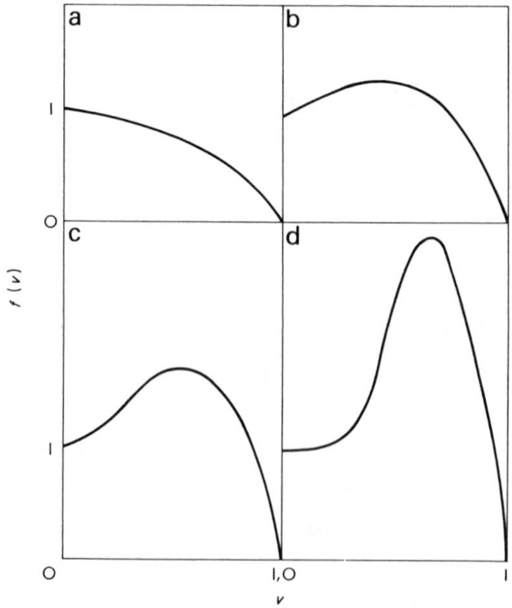

Figure 3 Possible forms of $f(v)$. a, monotonic; b, $f'(v)$ monotonic; c, $f(v)/v$ monotonic; d. $f(v)/v$ not monotonic

'O.K. show us then,' said k. 'Use your last example of diffusion and reaction:

$$\frac{d^2v}{dy^2} + \phi^2 f(v) = 0 \quad \text{in} \quad |y| < 1$$

and let's suppose the solution is symmetrical so:

$$\frac{dv}{dy} = 0, \quad y = 0,$$

$$v = 0, \quad y = 1$$

'Very well,' replied π, 'but first I must say a little more about the function $f(v)$. You notice that by definition it is zero when $v = 1$ and is 1 when $v = 0$; what's more it is positive in $0 < v < 1$. It follows that v is positive in $0 \leq y \leq 1$ and has its maximum at $y = 0$.'

'How so?' asked k.

'Elementary, my dear Boltzmann,' said π. 'If f is positive $v'' = d^2v/dy^2 = -\phi^2 f(v)$ is negative. So:

$$v'(y) - v'(0) = \int_0^y v''(\bar{y})\,d\bar{y} < 0$$

and, since $v'(0) = 0$, $v'(y)$ is always negative except at $y = 0$ where it is zero. This $v(y)$ decreases as y increases, so $v(y)$ must be greatest at $y = 0$ and least at $y = 1$. But $v(1) = 0$ so, if it is least there, it must be positive elsewhere. Notice that we have only used the fact that $f(v)$ is positive in $0 \leq v \leq 1$.

'But how do you know v can't be greater than 1 and so make $f(v)$ negative,' asked Re.

'By the same kind of argument, for, as you say, $f(v) < 0$ if $v > 1$. Suppose $v(0) > 1$, then $v''(0) > 0$ and there is a minimum of v at $y = 0$. Moreover:

$$v'(y) = v'(y) - v'(0) = \int_0^y v''(\bar{y})\,d\bar{y} > 0,$$

so that $v(y)$ increases as y increases and it can never come down to zero to satisfy the boundary condition.'

'Anyway it's obvious on physical grounds,' k was pleased to be one up on π. 'When you derived the equation, you substituted $c = c_0(1 - v)$ for the concentration and, since a concentration cannot be negative, v cannot be greater than 1.'

'Certainly that's convincing enough,' replied π, 'but it's no bad thing to get it out of the mathematics. For one thing, it would imply a serious mistake in the model building if the two arguments didn't agree. For another, the physical argument might not be so obvious in another context. What's more the mathematical argument may be developed for other purposes. For example, let us ask whether the solution for a given value of ϕ^2 is unique. The standard way of doing this is to suppose that there are two solutions v_1 and v_2 and then prove that their difference vanishes; that is, there are not really two distinct solutions but only one. If v_1 and v_2 are both solutions, both of them satisfy the equations:

$$v_1'' = -\phi^2 f(v_1) \quad \text{and} \quad v_2'' = -\phi^2 f(v_2),$$

and both satisfy the boundary conditions

$$v_1'(0) = v_2'(0) = v_1(1) = v_2(1) = 0.$$

If we take the equation for v_1 and multiply it by v_2 and take the equation for v_2 and multiply it by v_1 and then subtract we have:

$$v_2 v_1'' - v_1 v_2'' = -\phi^2[v_2\, f(v_1) - v_1\, f(v_2)] \tag{1}$$

Now integrate both sides between $y = 0$ and $y = 1$. For the left hand side we have:

$$\int_0^1 [v_2 v_1'' - v_1 v_2'']\,dy = [v_2 v_1{}' - v_1 v_2{}']_0^1$$

But this is zero, because $v_1{}'$ and $v_2{}'$ are both zero at $y = 0$ and v_1 and v_2 are zero at $y = 1$. It follows that the integral of the right hand side of (1) is also zero and, since $\phi^2 > 0$:

$$\int_0^1 [v_2 f(v_1) - v_1 f(v_2)]\,dy = 0$$

The integrand may be written:

$$[(f(v_1)/v_1) - (f(v_2)/v_2)]v_1 v_2$$

and if the integral is to vanish the integrand must either vanish everywhere (in which case $v_1 \equiv v_2$ and there really is only one solution) or it must change sign so that the integral over the negative parts equals the integral over the positive parts and the whole integral balances out to zero. Now $F(v) = f(v)/v$ has the geometrical interpretation of being the slope of the line joining the origin to a typical point $(v, f(v))$ on the $f(v)$ curve. When $v = 0$, $F(0)$ is infinite since this line is vertical; when $v = 1$, $F(1) = 0$ since $f(1) = 0$. Let us suppose $f(v)$ is monotonically decreasing* then $F(v)$ also decreases monotonically from infinity to zero as v goes from 0 to 1. Then $F(v_1) - F(v_2)$ can only change sign if v_1 is sometimes greater than v_2 and sometimes less, that is, the two curves $v_1(y)$ and $v_2(y)$ are intertwined. Let's see whether that's possible.

'If they are intertwined we can take $y = \eta$ to be the first point (i.e. least value of y) for which $v_1 = v_2$. We lose no generality by calling v_1 the solution which has the greater value at $y = 0$ (i.e. $v_1(0) > v_2(0)$). Then where they meet $-v_1'(\eta)$ must be greater than $-v_2'(\eta)$, since the situation must look like *Figure 4*. Now

$$\begin{aligned} -v_1'(\eta) &= -v_1'(\eta) + v_1'(0) \\ &= -\int_0^\eta v_1''(\bar{y})\,d\bar{y} \\ &= +\phi^2 \int_0^\eta f(v_1)\,d\bar{y} \end{aligned}$$

and

$$-v_2'(\eta) = \phi^2 \int_0^\eta f(v_2)\,d\bar{y}.$$

Since we have made the stronger assumption that $f(v)$ is monotonically decreasing (this implies, but is not implied by, the monotonicity of $F(v)$) then $v_1 > v_2$ in

* This is not the most general condition but it illustrates the point easily. cf. vol. 2 of reference 15.

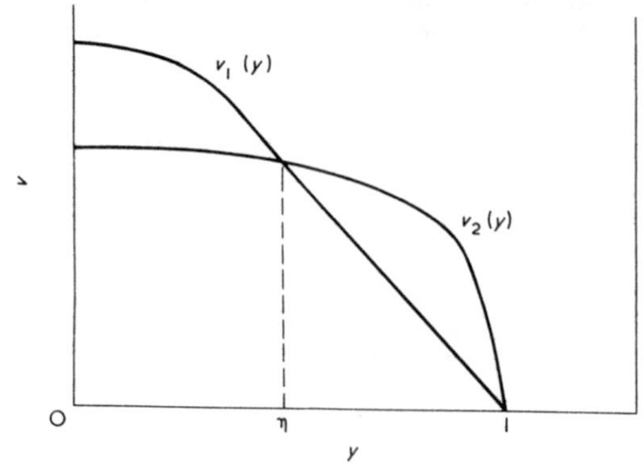

Figure 4 Disposition of two intertwined solutions

$0 \le y < \eta$ implies that $f(v_1) < f(v_2)$ and hence that $-v_2'(\eta) > -v_1'(\eta)$. But this contradicts the geometrical necessity that $-v_1'(\eta) > v_2'(\eta)$. So if $f(v)$ is monotonically decreasing the curves cannot intertwine.

'You're making it very easy for yourself with all these restrictions,' said k. 'Is that really cricket?'

'By all means,' π replied. 'At this stage of the game it isn't a bad thing to make simplifying assumptions for we are still feeling out the problem. What is bad is to forget you've made 'em and so delude yourself into thinking you know more than you really do. Assumptions should be tagged so that you know that you have to come back and try to relax them.'

'That's reasonable enough.' Re looked as if he wanted to get on with the job of solving the equations. 'We now know that $v(y)$ is positive and that the solution is certainly unique if $f(v)$ is monotonic. Shouldn't we just slap it on the computer? At this point in time?' he added knowing that the solecism would get π's goat.

'Not quite yet, *if* you please.' π was using an exaggerated politeness to restrain his wrath at the Nixonian use of English. 'We know that the solutions are monotonic, decreasing as we move outwards from $y = 0$ to $y = 1$, but the equations still have boundary conditions at two different points so that if we move outwards from $y = 0$ where we know $v'(0) = 0$ and make up for our ignorance by assuming a value for $v(0) = v_0$, how do we know that we shall hit $v(1) = 0$ exactly at $y = 1$?'

'By trial and error, of course,' said Re.

'You could simultaneously integrate an equation for $\partial v(y)/\partial v_0$,' added k. 'Then, when you reach $y = 1$ you have an idea of how sensitive $v(1)$ is to changes in v_0. In fact you can iterate by changing v_0 to $v_0 - v(1)/[\partial v(1)/\partial v(0)]$—Newton–Raphson, you know,' he added rather patronizingly, seeming to assume that anything with the name of Newton attached was the peculium of the physicist.'

'You're still generating a lot of incinerator-fodder, when you could make every calculation count,' said π. 'Why not use the second non-dimensional form with:

$$z = \phi y = \phi x/L = x\sqrt{k(T_0)/D}\ ?$$

Start by assuming some value of v_0 in the interval $(0, 1)$; integrate the equation $d^2v/dz^2 = f(v)$ from $y = 0$ with $v(0) = v_0$ and $v'(0) = 0$ until $v(z) = 0$. Since $y = 1$ when $v = 0$, z must equal ϕ there. In other words, you determine the value of ϕ for which v_0 happens to be the right value of $v(0)$. This is better than hunting for the v_0 that corresponds to a preordained ϕ. Always be prepared to interchange the things you assume with those you have to determine.'

'Is this what you call the just moulding of models?' k asked.

'Yes; there are many more tricks to the trade—as will be discovered to you if you come to my course.' π was unable to resist the commercial. 'Maximum principles, phase planes and isoclines, *a priori* bounds, group theoretic methods, quick and dirty calculations, generating functions, variational bounds, perturbations both regular and singular, integral methods even the lowly notion of integration by parts—I treat 'em all with great care and distinction.'

'A veritable "joy of mathematical modelling", by the sound of it.'

But not only was π's complacency proof against the stone of k's sarcasm, he positively caught it as it were a ball and ran with it. 'Actually I often call it the foreplay of applied mathematics because it requires a similar tact and tentativeness, a corresponding restraint and delicate imaginativeness that are characteristic alike of lover, poet and mathematician. Oh, and freedom from solemnity,' he added, pulled back from the brink of betraying his own words by the memory of Lewis' remarks*.' Morse and Feschbach[16] or Courant and Hilbert[17] at this stage would be like having Freud, Kraft-Ebbing and Havelock Ellis on the bed-table.'

'Good grief!' muttered Re under his breath, looking across at k who had raised his eyes to heaven in acknowledgement of the fact that this one had backfired on him. 'But tell us about the presentation of models. That's just the illustrations, I suppose.'

'Oh, heavens, not just the illustrations. It's ...' π faltered as it suddenly dawned on him that his prose was being compared with Comfort's and his figures with.... He went on hastily. 'No, figures are of great importance, of course, and may at some times be drawn with accuracy and at others distorted for clarity, or even dramatic, effect. Synopsis should be cultivated so that as much as possible the model is seen steadily and seen whole. This is one of the chief virtues of catastrophe theory which, as Thom says,[18] "is not a mathematical theory but a 'body of ideas'". I would say it's a way of looking at things and a great part of its value lies in helping people grasp the situation more completely. But there are other interesting questions to examine such as the use of observable, rather than intrinsic, parameters. In the diffusion problem we were

*'All my life a ludicrous and portentous' solemnization of sex has been going on ... And the psychologists have so bedevilled us with infinite importance of complete sexual adjustment and the all but impossibility of achieving it, that I could believe that some young couples now go to it with the complete works of Freud, Kraft–Ebbing, Havelock Ellis and Dr. Stopes spread out on bed tables around them. Cheery old Ovid, who never either ignored a molehill or made a mountain of it, would be more to the point. We have reached the stage at which nothing is more needed than a roar of old-fashioned laughter." Lewis, C. S. 'The Four Loves' (p. 113) G. Bles, London, 1960

discussing $\phi^2 = L^2 k(T_0)/D$ is an intrinsic parameter, because $k(T_0)$ is the intrinsic reaction rate in the absence of diffusion limitation. You can't measure $k(T_0)$ unless D is very large (or ϕ is small) but you can't tell if ϕ is small until you've measured ϕ. Of course there are times when the intrinsic rate can be measured but it is important to find out what is measured when diffusion is affecting the measurement. From the way we have defined it $k(T_0)f(v(y))$ is the actual rate of reaction at a point y so:

$$k_{\text{obs}} = k(T_0) \int_0^1 f(v(y))\,\mathrm{d}y$$

is the total rate of reaction. If D is very large ϕ is almost zero and the solution of $v'' + \phi^2 f(v) = 0$ is very close to the solution of the equation when $\phi = 0$, namely $v(y) \equiv 0$. But $f(0) = 1$ so that when $\phi \to 0$ we do measure the intrinsic rate, since

$$k(T_0) \int_0^1 f(0)\,\mathrm{d}y = k(T_0).$$

For any value of ϕ let $\eta = \eta(\phi)$ be the value of $\int_0^1 f(v(y))\,\mathrm{d}y$. We get η by solving the equation for $v(y)$ and evaluating the integral so it is a function of ϕ (*Figure 5*). Then the observed rate constant k_{obs} and the intrinsic k are related by $k_{\text{obs}} = \eta k$. If then we define $\Phi^2 = L^2 k_{\text{obs}}/D$ as an observable version of $\phi^2 = L^2 k/D$, we have $\Phi^2 = \eta(\phi)\phi^2$. If $\phi^2\eta(\phi)$ is a monotonic function of ϕ, then we can express η as a function of ϕ. For example, if $f(v) = 1 - v$, $\Phi^2 = \phi \tanh\phi$ which defines an inverse function $\phi(\Phi)$ and then $\eta(\Phi) = \Phi^2/[\phi(\Phi)]^2$. For practical use $\eta(\Phi)$ is the better function to tabulate since it will show how much diffusion limitation there is (η close to 1 means little effect of diffusion $\eta \ll 1$ means severe-diffusion limitation) in terms of an observable parameter.'

'And I never knew you cared,' murmured Re, rising and picking up his books.

'But how else can we defend or prove them when they are put to the test*.' Ignoring Re's cheek, π dropped unknowingly into the allusion the two had started out with and k promptly took it up.

'Ah, yes,' he said, 'so they are to be called not only mathematicians, engineers or physicists, but are worthy of a higher name—the modest and befitting title of natural philosophers*.'

Soc. And now the play is played out; and of rhetoric enough. Go and tell Lysias that to the fountain and school of the Nymphs we went down and were bidden by them to convey a message to him and to all other composers of speeches... to all of them we are to say that if their compositions are based on knowledge of the truth, and they can defend or prove them, when they are put to the test, by spoken arguments, which leave their writings poor by comparison of them, then they are to be called not only poets, orators, legislators, but are worthy of a higher name, befitting the serious pursuit of their life.

Phaedr. What name would you assign to them?

Soc. Wise, I may not call them; for that is a great name which belongs to God alone, —lovers of wisdom is their modest and befitting title.

Phaedrus 278, (Jowett's translation)

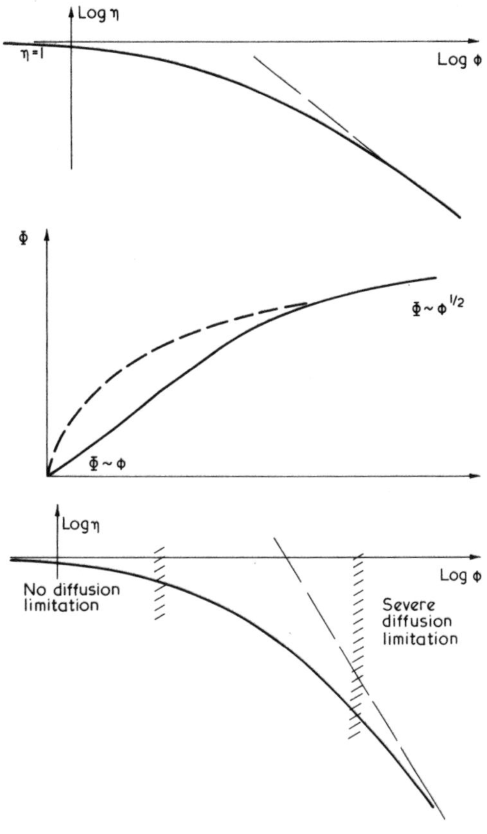

Figure 5 Effectiveness of a catalyst in terms of Thiele modulus (top), observable modulus (bottom) and relation between moduli

References

1 Plato. Phaedrus, 230
2 ibid., 228
3 Maynard Smith, J. 'Models in Ecology', Cambridge University Press, London, 1974
4 May, R. M. *Nature* 1976, **261,** 459
5 Li T-Y. and Yorke, J. A. *Amer. Math. Soc. Mon.* 1975, **82,** 985
6 May, R. M. and Oster, G. F. *Amer. Nat.* 1976, **110,** 573
7 Taylor, G. I. *Proc. Roy. Soc.* 1953, **A219,** 186
8 Aris, R. *Proc. Roy. Soc.* 1956, **A235,** 67
9 Gill, W. N., Ananthakrisnan, V. and Barduhn, H. J. *AIChE J.* 1965, **11,** 1063
10 Gill, W. N. *Proc. Roy. Soc.* 1967, **A298,** 335
11 Duhem, P. 'The aim and structure of physical theory'. (Trs. P. P. Wiener. Princeton University Press, Princeton, 1954. Original edition, 1906
12 Echecrates to Phaedo in Phaedo 57B.
13 Segel, L. A. *SIAM Rev.* 1972, **14,** 547
14 Lin, C. C. and Segel, L. A. 'Mathematics applied to deterministic problems in the natural sciences.' Macmillan Publishing Co., New York, 1974
15 Aris, R. 'The mathematical theory of diffusion and reaction in permeable catalysts.' 2 vols. Clarendon Press, Oxford, 1976
16 Morse, P. M. and Feshbach, H. 'Methods of theoretical physics.' 2 vols. McGraw-Hill, New York, 1953
17 Courant, R. and Hilbert, D. 'Methods of mathematical physics' trs. of Methoden der mathematischen physik. 2 vols. Interscience. New York, 1953.
18 Thom, R. *SIAM Rev.* 1977, **19,** 189

THE JAIL OF SHAPE

RUTHERFORD ARIS

Department of Chemical Engineering and Materials Science,
University of Minnesota,
Minneapolis, MN 55455

(Received April 30, 1983; in final form June 22, 1983)

PROLOGUE

The College of Saints Jerome, Paula, and Eustochium, an ornament to one of the great universities of the western world, finds almost universal approbation among those who know of it. Theologians are so glad to find a Doctor of the Church honoured that they overlook his reputation for irrascibility, while psychiatrists are charmed to find a genuine subject of their studies so exalted. The watchers of gender rejoice that the balance is two to one in favour of the female, though a sociologist, whose learning had not quite reached the beginnings of words, was heard to rejoice in the celebration of masculine, feminine and neuter. In point of fact the place was founded, and does indeed conduct itself in complete disregard of differences of sex and age as a veritable beacon in a naughty world, enshrining forever the oldest and fullest meaning of the word 'man.'

It was within its Inner Quadrangle on an uncommonly warm afternoon that three friends gathered. They would not wish their ages or sexes to be revealed, so I will designate them by a single majuscule. One was an art historian, with a bent for mathematics, who admitted to the designation of a single B—of course he (or she) would have liked to have had it doubled, but modesty disallowed. The second, an engineering scientist, was not so inhibited and off her (or his) own bat suggested L; the third, a professional scribe holding a visiting fellowship at the college that year, was a great admirer of Tagliente and accepted T. for a siglum. L. and T. disposed themselves in deck chairs while the portly B, who had a rooted distrust of these devices, sprawled on the grass and continued a conversation that had evidently been begun earlier.

DIALOGUE

"You were saying L," said B, "before old Cholmondeley interrupted at lunch, that you almost never worked with quantities having dimensions. I thought you engineers always did things with appropriate units—meters and litres and the like. Aren't you always quarreling about Celsius and Fahrenheit and that sort of thing?"

"Well, of course, a lot of engineering has to be done with units and a lot of care has to be taken over the preservation of the standards. Otherwise you couldn't have something designed in Japan and built in India. You have to teach students to use 'em, and they use textbooks that have been rewritten in SI units, whatever they are. Some of my colleagues get very worked up about 'metrication' and all that, but in the kind of problem I'm interested in I seldom, if ever, need feet or lbs, let alone meters and kilograms. They're all completely arbitrary anyway."

"But I thought there was a human background to units," said B. "The foot is of the size of a large foot, the inch can be measured off with a segment of a finger, barley corns are barley corns and carats, carats and so on."

"Yes, there's a lot of colorful background to dimensions though the rationalists of the French Revolution tried to do away with all that by defining the meter as the umpteenth part of the length of the equator. Of course they could not determine that with perfect accuracy and they finished up with the length of an arm anyway. But these are very vague and haphazard relationships and have little more than anecdotal value."

"I hope you're not deprecating anecdotes," interrupted B. "When all your scientific abstractions are swept away by the next paradigm, anecdote will remain a mirror unto man to whom it says . . . ."

"Spare us, spare us. I wasn't downplaying the humane aspect, but merely making the point that fundamental units are necessarily arbitrary, and because they must be used in all situations, are peculiar to none. Tremendous care has to be taken in propagating the units from the primary standard in Paris, through secondary and tertiary standards to the level of everyday life. But in the problems of academic, or at any rate theoretical, engineering science the magnitude of a quantity needs to be defined in the terms of the magnitudes of the particular situation. A fine powder may not be fine or small with respect to the molecules that are reacting with it whereas it may indeed be small compared with the paddles that are moving it around and mixing it up."

"That's not so strange to me," said T. coming into the conversation for the first time. "The size of a letter is always specified in terms of pen widths. Whether it's a fifth of an inch or five inches high may depend on the use it's being put to—it's no good writing a book with five inch letters or a carving monument in a minute script—but it's the proportion that gives it its character. A Rustic capital may be nine pen widths in height, but a Caroline minuscule is usually more like four." And suiting the action to the word, T. pulled out a broad felt-tipped pen, took up a pad (for T never seemed to be without them) and started writing. (What he drew is shown here as Figure 1.)

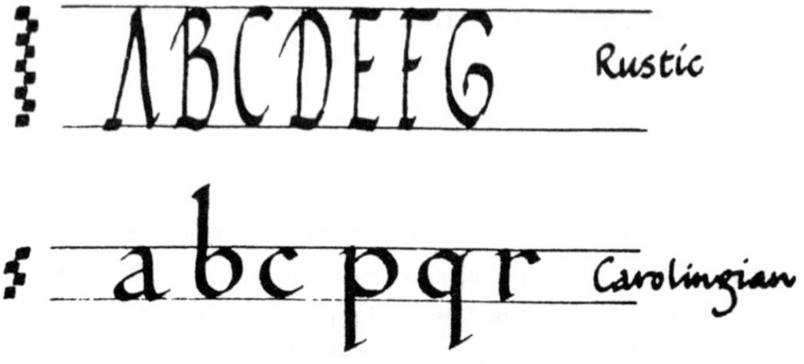

FIGURE 1 The different proportions of different scripts shown as dimensionless multiples of the pen width.

"There's much the same internal standard in a painting, of course," said B. "It's just a matter of getting the proportions right, or rather using the proportions for a particular purpose. I'll think of an example in a minute, but I'm not sure that I really followed L's notions except in a very trivial sense. Give us a really significant example and pitch it so that we can understand without too much of your beastly technological jargon."

"Well, just think of your automobile," L. began. "If it has been made in the last few years it has a catalytic converter in the exhaust system."

"If you say so," said B., who affected an immense ignorance of things mechanical. "I suppose it's the source of some of the stinks I get from my new machine. Does it help the mileage."

"Heavens no" said T., who clearly was much more aware of the banalities of modern living, or at the least was less shy in admitting to knowledge of them. "Nothing helps the mileage more than a change of dimensions. They measure them now in EPA miles, don't y'know—a euphemism for kilometers by the look of it."

"The catalytic converter's for cutting down the pollution, of course," said L. resuming his discourse. "I only referred to it because it's the commonest of all reactors these days and illustrates the point I want to make with a non-trivial case such as you wanted. All you need to know is that it's a heavy metal container filled with pellets through which the exhaust passes. The exhaust as it comes from the engine contains a greater percentage of pollutants than can be tolerated by EPA standards and these have to be removed by reaction. To keep it simple just think of carbon monoxide: this is a poison that has to be removed by converting it to carbon dioxide. Oxygen, from air, is present in the exhaust and the catalyst is needed to make the oxidation reaction that changes the carbon monoxide into dioxide go at a reasonable rate. If you don't mind I'll use the abbreviation CO for carbon monoxide—it happens to be its chemical formula."

"Now the catalyst pellets are porous and the reaction itself only takes place inside the pores when a CO molecule diffuses in from the mouth of the pore and sits down on the wall. These pores can be thought of as little cylindrical tubes about 100Å in diameter, but though that sounds small it is, of course, very many times bigger than the molecule. Sometimes the CO molecule alights on the surface very close to an oxygen molecule and to a platinum atom that has been put there in the preparation of the catalyst, and then it may be converted into the harmless carbon dioxide. The probability of this happening is proportional to the concentration of CO in the pore at a given position, and it's because this isn't what you might think it should be that a problem arises.

"You might think that the concentration of CO inside the pellet is the same as that on the outside, but not so. Because the CO has to diffuse into the pellet and it is there used up, its concentration falls off, and in a symmetrical pellet, such as a sphere, it is least at the center. In fact in one case it takes exactly the same form as the curve of a chain hanging freely between two points, the catenary. Here, let me draw it." And pulling out a pen, L. sketched a figure. (As it was equivalent to the small diagram in Figure 2 it is not reproduced specially.)

"Give us the equations," said B. who was particularly proud of having taken some advanced courses in what he quaintly called "the higher mathematics." "We're not unacquainted with the higher mathematics, you know," he added primly, and you could almost see the quotation marks round the words as he enunciated them.

pen-width variables

$$u = \frac{c}{c_s} \qquad x = \frac{r}{a}$$

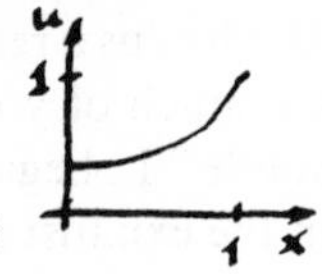

$$f(u) = \frac{R(c)}{R(c_s)} \qquad \varphi^2 = \frac{a^2 R(c_s)}{D c_s}$$

$$\boxed{\begin{array}{c}(x^2 u')' = \varphi^2 x^2 f(u) \\ u'(0) = 0 \;,\; u(1) = 1\end{array}}$$

$$\varphi^2 = 3\,\frac{(4/3)\pi a^3 R(c_s)}{4\pi a^2 D\,(c_s/a)}$$

$$\eta = \frac{\bar{R}}{\frac{4\pi}{3} a^3 R(c_s)} = 3\int_0^1 x^2 f(u(x))\,dx$$

FIGURE 2 The basic diffusion-reaction problem in the dimensionless variables.

"Oh, sure," said L., quickly listing the variables; r, the distance from this center; $c(r)$, the concentration of CO there; c_s, its concentration at the surface; D, the effective diffusion coefficient; $R(c)$, the rate of disappearance of CO per unit volume as a function of c; and, a, the radius of the pellet.

"Those are the variables and the parameters of the problem. Now think of a sphere of radius a; a balance over a very thin shell of inner radius r leads to the differential equation for $c(r)$

$$\frac{D}{r^2}\frac{d}{dr}\left(r^2\frac{dc}{dr}\right) = R(c)$$

and this has to be solved subject to the conditions

$$c = c_s \quad \text{at} \quad r = a,$$

and

$$\frac{dc}{dr} = o \quad \text{at} \quad r = o$$

B. grunted his approval in a knowing sort of way. "You don't have to know what it means," he told T. rather patronizingly. "It's just an ordinary differential equation that has to be solved for $c(r)$. I must confess I don't see how to solve it, though."

"Oh, you can't in general, except numerically," said L. who was rather surprised that B. had so far conceded his ignorance." But in any case, before you would try, you'd want to lick the equation into a much neater shape. You would think that there were at least four parameters – D, c_s, a, and one or more in R. But some of these can be used to make the variables dimensionless. These are your pen-widths, T."

L. quickly scribbled down the dimensionless concentration, $u = c/c_s$, and $x = r/a$, the dimensionless radial distance: he put $f(u) = R(c_s u)/R(c_s)$ and wrote the equation as

$$(x^2 u'(x))' = \phi^2 x^2 f(u)$$

$$u'(0) = 0, \qquad u(1) = 1 \quad \text{(cf. Figure 2)}$$

"The beautiful thing about this," L. continued, "is that there emerges a single parameter

$$\phi^2 = \frac{a^2 R(c_s)}{D c_s}$$

which, apart from the possibility that there are some parameters tucked away in $R(c_s)$, is the only parameter the problem needs. What's more, it has a clear physical meaning as the ratio of a reaction rate to a diffusion rate, for I can write it

$$\phi^2 = 3\frac{(4/3)\pi a^3 R(c_s)}{4\pi a^2 D(c_s/a)}$$

and the numerator is the volume of the sphere times a reaction rate per unit volume, while the denominator is the surface area times the diffusion coefficient times a concentration gradient."

"E.W. Thiele, an American chemical engineer who looked at this problem in the late '30's at about the same time as a German, Damköhler, and a Russian, Zeldowitch, went a step further and wrapped up the whole problem in a single number. What's needed is the ability to calculate the total rate of reaction, which is

$$\bar{R} = \int_o^a 4\pi r^2 R(c(r))\, dr = 4\pi a^3 R(c_s) \int_o^1 x^2 f(u(x))\, dx$$

Now, if there were no diffusional hindrance to the CO getting into the pellet, the concentration would be everywhere c_s and the total reaction would be the volume $(4/3)\pi a^3$ times $R(c_s)$. This is the natural measure of reaction rate and the ratio

$$\eta = \bar{R}/(4/3)\pi a^3 R(c_s) = 3\int_o^1 x^2 f(u(x))\, dx$$

is called the effectiveness factor."

"Why use two words when one would do?" asked B. "Surely 'effectiveness' would be quite unambiguous."

"I don't know," replied L. "It was a habit that stuck I suppose. Even Aris, who seems to fancy himself as a user of good English, didn't have the sense to drop the word 'factor' in a monograph he wrote on the subject a decade ago.[1] At least the phrase 'utilization factor' which appeared briefly in the literature has perished."

"I should hope so," said T. "How anyone could use the word 'utilize' is beyond me. But do I understand you aright that if your problem only has one parameter, ϕ, then η, the effectiveness, will depend only on ϕ?"

"Yes, though perhaps there will be further parameters tucked away inside $f(u)$. For example, if

$$R(c) = \frac{Kc}{1 + Kc}$$

$$f(u) = \frac{(1 + \kappa)u}{1 + \kappa u}, \qquad \kappa = Kc_s$$

In general the effectiveness η (known to mathematicians as a functional of the solution) gives a curve like this (and here L. sketched Figure 3). It must approach 1 as $\phi \to 0$ and it can be shown that $\phi\eta$ tends to be a constant as ϕ gets large. The curves labelled A, B, and C would be the concentration profiles in the pellet corresponding to the ϕ-values labelled A, B, and C on the ϕ, η-curve.

"Now the interesting thing is that the constant to which $\phi\eta$ tends are $\phi \to \infty$ depend only on the kinetics and the shape. The reason is that for high reactivity the concentration of reactant falls off exponentially quickly just inside the surface, and it matters less and less whether it's cylindrical or spherical or any other shape. It might as well be taking place in a plane slab. In the slab of half thickness a, the differential

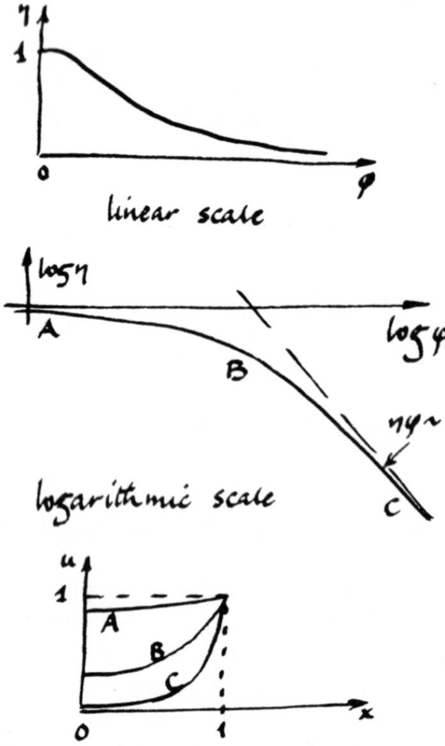

FIGURE 3 The relation of the effectiveness to the Thiele modulus.

equation is

$$u'' = \phi^2 f(u)$$

$$u'(0) = 0, \qquad u(1) = 1$$

and this can be integrated by quadratures."

"By quadratures," said T., sitting up. "That's a quaint phrase on modern lips. 'Capitalis quadrata' is an old name for Square or Roman capitals and 'quadrature' is literally 'making square.' What does it mean in your context?"

"I suppose it gets its meaning from the idea of making something square or rectanglar. You are really making an irregular area equivalent to the area of a rectangle which is easily calculated. Quadrature is just the process of calculating the area under a curve."

"Integration, don't you know," interrupted B., as if that explained everything to T. who had understood perfectly well anyway.

"Well, multiplying my equation by $2u'$ and *integrating*," said L., with heavy emphasis," gives

$$[u'(x)]^2 = 2\phi^2 \int_{u_0}^{u(x)} f(u)\,du = \phi^2[F(u; u_0)]^2$$

where u_0 is the dimensionless concentration at the center of the slab. We don't know the value of u_0 yet, but it is between zero and one. Integrating again and using the boundary condition $u(1) = 1$ gives

$$\phi x = \int_{u_0}^{u(x)} \frac{du}{F(u; u_0)}$$

and

$$\phi = \int_{u_0}^{1} \frac{du}{F(u; u_0)}$$

The flux per unit area at the surface is

$$D\left(\frac{dc}{dr}\right)_{r=a} = D\frac{c_s}{a}\left(\frac{du}{dx}\right)_{x=1} = D\frac{c_s}{a} = \phi F(1; u_0).$$

If the outside area is S_x and the pellet volume is V_p

$$\eta = \frac{S_x D(c_s/a)\phi F(1; u_0)}{V_p R(c_s)} = \frac{a S_x F(1; u_0)}{V_p \phi}$$

Now as $u_0 \to 0$, $\phi \to \infty$ because the reaction goes so fast that it completely exhausts the reactant, so asymptotically

$$\eta\phi \sim \frac{aS_x}{V_p} F(1; 0)$$

If instead of the radius or half-thickness, a, we use $[V_p/S_x F(1; 0)]$ to define the characteristic dimension, i.e. define

$$\phi^2 = \frac{V_p^2}{S_x^2} \frac{R^2(c_s)}{D} \bigg/ \int_0^{c_s} R(c)dc$$

then we have

$$\eta\phi \sim 1$$

for all shapes and kinetics.

"There is an important exception to this asymptotic behavior," went on L. somewhat intoxicated with the exuberance of his own verbosity, and that is when the boundary condition is not $c = c_s$ at the surface, but $k_c(c_f - c_s) = D(\partial c/\partial n)_s$. This simply says that there is a certain resistance to the transfer of carbon monoxide between the flowing gas, where its concentration is c_f, and the surface, where its concentration is c_s. In dimensionless terms this is $u'(1) = v$ where $v = k_c a/D$. For a first order reaction it's been known for a long time that the reciprocal of the internal effectiveness is a kind of resistance which is additive to the external resistance proportional to $1/v$, so that asymptotically

$$\frac{1}{\eta} \sim \Phi + \Phi^2/v.$$

But I proved the other day that the same formula holds good for any kinetics as $\Phi \to \infty$ and $v^{-1} = o(\Phi^{-1})$. Here, I'll show you." And, seizing T.'s pad, L. sketched what has been tidied up in Figure 4. It took about five minutes of handwaving, for L. was so pleased with himself at being able to recall it all that he didn't notice he had completely lost his listeners.

"It is not surprising therefore," he went on in the full flood of his eloquence, "that when the activity varies through the pellet a special case is obtained when the activity falls to zero at the surface, for that's really like having an external resistance. Varma, working with Wang and Morbidelli[2] showed that a general result could be obtained by suitably modifying v to include both the external resistance and that of the dead exterior region. Yortsos and Tsotsis[3] extended this result using the Liouville-Green or WKBJ method, when the activity varies but does not vanish within the pellet. In the case where the activity becomes zero at the surface in a known and fixed way (i.e. quadratically or cubically) a shape normalization can be done. But a general normalization is not possible. So you see shape's a tricky thing in chemical engineering."

Not even a snore greeted L.'s conclusion, which B. seemed to regard as profoundly trivial, and T. looked completely lost. B. turned to T. and said,

"Why don't you tell us about shape in letters, T.?"

"Hang on just one more minute," said L. hastily, "while I tell you about multiple first order reactions."

"Matrices, I suppose," said B., who when L. agreed with him was suffused with such complacency that he sat up in his best approximation to the lotus position and stayed wide awake the rest of the time.

"Yes, the extension to any set of first order reactions is simple; you just replace the concentration by a vector of concentrations $\underset{\sim}{u}$ and the diffusion coefficient and rate constant by matrices $\underset{\sim}{D}$ and $\underset{\sim}{K}$. There are good physical reasons why the matrix $\underset{\sim}{D}$ must have an inverse, and we can write $\underset{\sim}{\Psi}^2 = \underset{\sim}{D}^{-1}\underset{\sim}{K}$ to give

$$\nabla^2 \underset{\sim}{u} = \underset{\sim}{\Psi}^2 \underset{\sim}{u} \quad \text{in} \quad \Omega$$

$$\underset{\sim}{u} = \underset{\sim}{u}_s \quad \text{in} \quad \partial\Omega$$

Slab - external transfer resistance

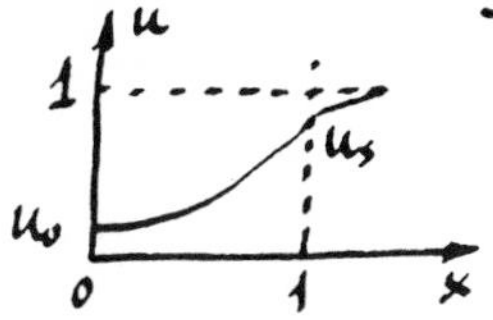

$$u'' = \varphi^2 f(u) \; ; \quad u'(0) = 0 \; , \; u'(1) + \nu u(1) = \nu$$

gives $$[u'(x)]^2 = \varphi^2 F^2(u(x);u_0) = 2\varphi^2 \int_{u_0}^{u(x)} f(u)\,du$$

and hence

$$\varphi = \int_{u_0}^{u_s} du / F(u;u_0)$$

$$\varphi\eta = F(u_s;u_0) = \frac{\nu}{\varphi}(1-u_s)$$

Also $$F(u_s;u_0) = F(1;u_0) - (1-u_s)/F(1;u_0)$$

by Taylor's thm, whence

$$F - \frac{1-u_s}{F} = \frac{\nu}{\varphi}(1-u_s)$$

and $$\frac{1}{\eta} = \frac{\varphi^2}{\nu F^2} + \frac{\varphi}{F} = \frac{\Phi^2}{\nu} + \Phi$$

FIGURE 4 The general asymptotic problem with external resistance.

Now in most cases we can diagonalize $\underset{\sim}{\Psi}$ by finding a matrix $\underset{\sim}{L}$ such that

$$\underset{\sim}{L}^{-1}\underset{\sim}{\Psi}^2\underset{\sim}{L} = \operatorname{diag}(\psi_1^2, \psi_2^2, \ldots, \psi_n^2)$$

where what is called the definiteness of $\underset{\sim}{\Psi}$ ensures that the ψ_i are real though, if all the species are included, $\psi_1 = 0$. If

$$\underset{\sim}{v} = \underset{\sim}{L}^{-1}\underset{\sim}{u}, \qquad \underset{\sim}{v}_s = \underset{\sim}{L}^{-1}\underset{\sim}{u}_s$$

and $V(\rho;\phi)$ is the solution of the scalar problem

$$\nabla^2 V = \phi^2 V \quad \text{in} \quad \Omega$$

$$V = 1 \quad \text{on} \quad \partial\Omega$$

with

$$\eta(\phi) = \frac{1}{V_p} \iiint V(\rho; \phi)\, dV$$

then

$$\underset{\sim}{v} = \text{diag}(V(\psi_1), V(\psi_2), \ldots, V(\psi_n))\underset{\sim}{v}_s = \underset{\sim}{\Delta}\underset{\sim}{v}_s \text{(say)}.$$

Thus,

$$\underset{\sim}{u} = \underset{\sim}{L}\underset{\sim}{v} = \underset{\sim}{L}\underset{\sim}{\Delta}\underset{\sim}{L}^{-1}\underset{\sim}{u}_s$$

and the average rate of reaction is

$$\frac{1}{V_p} \iiint \underset{\sim}{K}\underset{\sim}{u}\, dV = \underset{\sim}{K}\underset{\sim}{L}\underset{\sim}{E}\underset{\sim}{L}^{-1}\underset{\sim}{u}_s = \underset{\sim}{H}\underset{\sim}{K}\underset{\sim}{u}_s$$

where

$$\underset{\sim}{E} = \text{diag}(\eta(\psi_1), \ldots, \eta(\psi_n)).$$

"That's a truly amazing result," concluded L. triumphantly. "It shows that the first-order irreversible effectivenesses can be used in the most complicated network of reversible first-order reactions. Indeed any formula that can be written in terms of effectivenesses is shape invariant in first-order systems. Put another way, we can relate the results of a whole class of problems to the result of the simplest problem of its class. The problems for different shapes have a common bond and as it were, a certain connection between themselves."

"Sounds to me as if you've got off on Cicero's famous phrase: 'quasi cognitione quadam inter se continentur,'" said T.

"What's that?" said L., thoroughly confused.

"Oh nothing," said T. quickly, fearful lest B. should take it up with his usual pomposity. "Just a tag I happened to think of. But let me tell you about shape relations in script. Albrecht Dürer's treatise on applied geometry "Underweysung der Messung mit dem Zirckel und Richtscheyt" first published in 1525, has a famous chapter known by its translated title "Of the Just Shaping of Letters." His diagram capitals are well-known and illustrate the way in which they fit into the square and indicate the curvature of various parts by inscribed circles.

"He also gave a prescription for Gothic or, as he called them, "text" letters. "It was formerly the custom so to write," he says, for by 1525 the humanistic script had gained the ascendency over the Gothic script. He bases everything on the letter "*i*" for this is the dominant component of Gothic script, and in a word like "minimum" (Figure 5a) gives it the almost unreadable regularity of a picket fence. (T. was busy all the time with his pen while he was talking and some of his demonstrations are collected in Figures 5 and 6). His 'quadrate' *i* is five pen widths tall, but the bottom of the top square and the top of the bottom one are divided into three parts on which like squares, but with

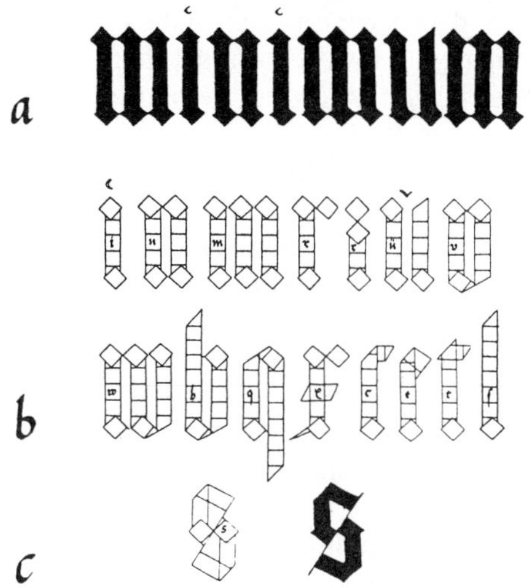

FIGURE 5 a. The word 'minimum' in Gothic script.
b. Durer's canons for Gothic.
c. The construction of Gothic *s*.

vertical diagonals, respectively stand on the left-most third mark and depend from the rightmost (Figure 5b).

"The letters *m*, *n*, *u*, *v*, *w* are combinations of *i* with some indication of joining at the bottom of *v* and *w*. The letter *r* has either two diagonal squares at the top or, more usually, a second jutting out to the right; *b* ascends a further three squares while *q*, its image under a rotation of π, descends equally. The tall letters *b*, *f*, *h*, *k*, *l*, and long *s* are all eight squares high (Figure 5b). The short *s* gets his most elaborate directions (cf. Figure 5c): 'Curved, or short *S*,'you shall make on this wise. At the middle height of the letter, let there be set, close to one another, their angles touching, two oblique squares; from the near square draw a broad vertical limb to the height of the letter; and in the same fashion, from the farther square let one fall downwards—just as you constructed I top and bottom. Next cut off both these limbs, one at top and one at bottom, by diagonals, in such fashion that the sharp tips of both may be on the side near the middle. Then let there be drawn two broad limbs—namely, from the upper, to the right, and downwards; and in like manner; from the lower, upwards, and to the left; of the breadth of the limb, above and below, but let them be produced no further than the breadth of the distance between the limbs: then draw a diagonal downwards, from right to left, which shall cut off both oblique limbs. To it also you must produce the sides of the squares set in the midst.

"That's all very easy for Gothic," said B., "but how can you hope to do it for a more curved script such as the humanistic hands that superseded Gothic?"

"Let me go back to an earlier script," answered T. "After all the Italian humanists took as their model the 9th century hands of the Carolingian period, though they thought that they were reaching back to antiquity. It's what we would call a good round hand and in fact is the basis of all our modern typefaces, for it was adopted by the first printers and has persisted with various modifications ever since. As it was developed over the 10th and 11th centuries at centers like Winchester it acquired an outstanding grace and dignity and was the model that Edward Johnston took in his reform of manuscript lettering at the turn of the century. (Figure 6a shows three lines from a Winchester psalter).

"The thick and thin comes from the broad cut-nib and the direction in which it is moving and not from any variation of pressure—after all they used quills not steel nibs in those days, and its only with copperplate that you find pressure being used to make thick and thin. It is more difficult to assess how much variation of pen angle there was. In the traditional scripts there seems to have been very little, if any, and the beginner is usually taught to keep the angle of the pen constant. This is a wise introductory discipline but, as all good teachers will explain, no rule is inflexible and ultimately the scribe will move with such freedom that the letter on the page is but the record of where a three-dimensional dance of the pen has touched down on the two-dimensional space of the membrane. Then indeed there may be discovered variations of pen angle and subtleties of movement that the undisciplined hand and eye could not accomplish however great their virtuosity at pen twirling. There is a preface writer in New York

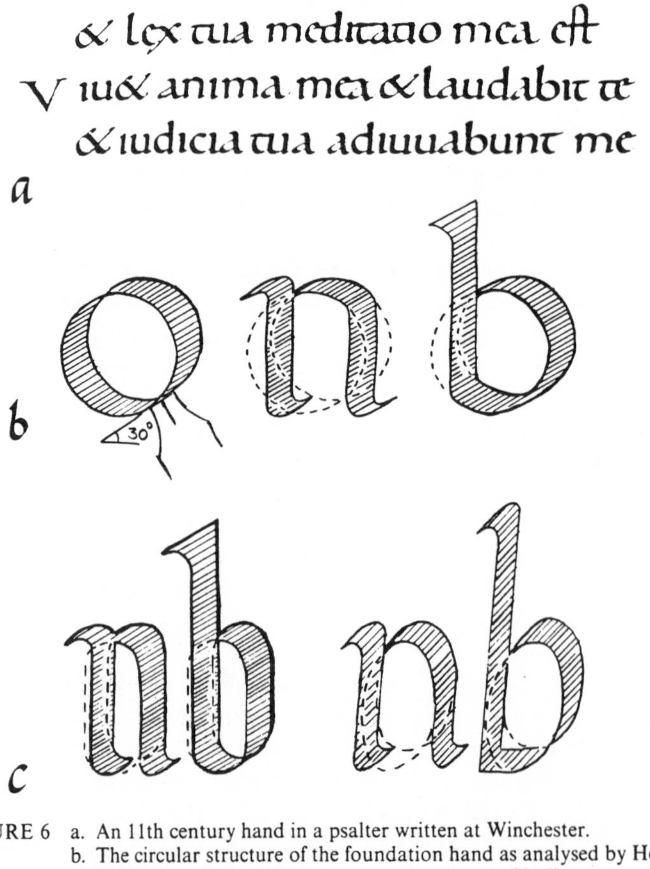

FIGURE 6 a. An 11th century hand in a psalter written at Winchester.
b. The circular structure of the foundation hand as analysed by Hechle.
c. Hechle's analysis of the compressed hand and of Italic.

who thinks a new age of freedom from the domination of the Johnstonians has been achieved by learning to twiddle from the start, but his case is far from made."

"I say, my dear fellow, don't get all worked up about that," interposed B., who, of course, knew that the word 'fellow' was of common gender. "Just tell us how you analyze the Carolingian minuscule on the principles of its shape. Do you take the *i* as the basic unit?"

"No, not the *i* but the *o*," T. replied, fetching a curious object from a side pocket and turning over a new leaf of his pad. The object turned out to be red and green felt-tip pens bound together, with which T. drew a circular *o* keeping the angle between the red tip on the left and the green on the right at 30° to the horizontal. "The pen of course fills in the ink between these lines which represent the track of the two edges of the nib. You can see quite clearly that the thinnest parts of the letter are at 11 and 5 o'clock when my pen angle is 30° (cf. Figure 6b).

"But what is even more interesting is the relationship of the letters to each other and to the *o*. Here I am following the analysis of Ann Hechle, one of the foremost calligraphers in Britain today and a fine teacher as well. She emphasises of course that it is just a guide to the underlying construction and not a set of rigid rules. I find it a fascinating analysis and the clearest explanation of the visual coinherence of the different alphabets. The round hand, which is a good foundational hand to learn, is based on the circular *o* above four and a half pen widths high and drawn with the pen at 30° to the horizontal. How the vertical ascender of *ab* or descender of *ap* is not just tangential to the *o* which is part of each of these two letters, rather the right side of the vertical stroke goes through the thinnest part of the letter at 11 o'clock. The upper part of the bow of the *b*, running clockwise from 11 to 2 thus starts to thicken immediately. If the vertical were much farther to the left the arch would look weak; if to the right it would be cramped and the round look would be lost. In the *n*, for example, the second stroke which begins the arch must start with the left hand edge of the nib at the left hand side of the first vertical stroke. Then the arch begins to thicken immediately and is strong, otherwise it looks stubby or weak. The other letters are related to the *o* in their several ways." And here T. sketched several letters some of which are shown in Figure 6b.

"That relates the different letters to one another through their common relationships to a single letter alright," agreed B. "But what about other alphabets?"

"Hechle's insight works here, too," T. insisted. "For example if the elongated 0 is taken as the base shape, a compressed round hand comes about. This seems to be something of the same relationship as L. had between his first order systems and the simple irreversible reaction, though I didn't fully understand all L. had to say."

"It was just a question of matrices," said B. smugly, though in fact he (or she) had understood no more of the mathematics than had T.

"Let me show you the compressed round hand which is based on the squashed circle. Again the inner side of a vertical goes through the thinnest part of the basic 0, and a more condensed upright hand results. Then the italic is based on an ellipse with a slightly greater pen angle of 45°. This time the ascenders and descenders are slightly sloped and are tangential at the outer edge rather than the inner edge going through the thinnest point. This gives the italic arch its characteristic leap and leaves scope for variation by altering the inclination of the ellipse and the vertical (Figure 6c). Sheila

Waters, one of the finest calligraphers practicing in the USA, has developed a Caroline minuscule based on a slightly flattened *o*—a grapefruit rather than an orange shape—that has great coherence and grace. She has used it in an exquisite manuscript of Dylan Thomas' "Under Milk Wood" which she wrote at the commission of an English collector. It can be analysed on similar lines.

"It is this freedom within a very disciplined framework that gives the art and craft of the scribe its perpetual fascination, and it seems to me has an analogy with the art and craft of mathematical modelling. Sometime I'd like to pursue the further analogy between the choice of script and evolution of a work of lettering art and strategy of the approach to a mathematical problem, large or small. It would be interesting to get two virtuosos such as Donald Jackson and John Hammersley talking to one another."

"Yes, that would be fascinating," agreed L., who seemed to be ready to explore the idea further when B. interrupted.

"At the risk of being pompous (a possibility that B. had never worried about before and was scarcely likely to start worrying about now) I might say that this reminds me of Sackville-West's lines:

'All craftsmen share a knowledge
They have held reality down fluttering to a bench;
Cut wood to their own purposes; compelled
The growth of pattern with the patient shuttle;
Drained acres to a trench.'"

"That reminds me, it must be time for afternoon tea," said L. with Philistine brutality. And so they departed the quad.

EPILOGUE

This conversation is reported as best I was able to get it down and is offered to the reader in the hope that it may stimulate discussion of the way we think about shape in engineering contexts as well as in everyday life. There is nothing particularly original in the engineering science here presented, though I do not recall having seen the general normalization presented in Figure 4 in quite that form before. The figures have been redrawn from the scraps of paper that I pounced on when B., L. and T. went for their afternoon sandwich.

The title is taken from the continuation of the poem quoted by B. at the end of the conversation:

They have ignored the subtle
Release of spirit from the jail of shape

The other references, without which no scientific paper, even in this guise, would pass muster are listed below with numbers corresponding to those I have inserted in the dialogue.

Finally I would like to thank Ann Hechle for permitting me to describe her analysis of letter forms and Jarda Ulbrecht for his editorial tolerance.

REFERENCES

1. Aris, R., *The Mathematical Theory of Diffusion and Reaction in Permeable Catalysts.* Clarendon Press, Oxford, 1975. The history of the shape factor and further references can be found there.
2. Varma, A., and Wang, J.B., "On Shape Normalization for Non-uniformly Active Catalyst Pellets", *Chem. Eng. Sci.*, **33,** 1549 (1978); **35,** 613 (1980), and Varma, A., and Morbidelli, M., *Chem. Eng. Sci.*, **38,** 297 (1982).
3. Yortsos, Y.C., and Tsotsis, T.T., "Asymptotic Behavior of the Effectiveness Factor for Variable Activity Catalysts", *Chem. Eng. Sci.*, **37,** 237 (1982). "On the relationship between the effectiveness factors for the Robin and Dirichlet problem for a catalyst with variable catalyst activity", *Chem. Eng. Sci.*, **36,** 1734 (1981).

THE MERE NOTION OF A MODEL

RUTHERFORD ARIS AND MISCHA PENN

University of Minnesota
Minneapolis, Minnesota 55455

Communicated by X. J. R. Avula

(*Received for publication April 1980*)

Abstract—Some of the ideas nascent in the notion of a model are here explored in a preliminary way. In particular we ask what the special insight afforded by the mathematical model may be and suggest that the notion of craftsmanship in mathematical modelling can provide a via media between the formalistic and subjective extremes, at once doing justice to the modeller's mental activity and providing a basis for dialogue and discussion.

We do not use the word "mere" in the sense of that notion of a model at which the hard-core empiricist would throw up his hands. For the word "mere" is patient of almost opposite meanings [1]. On the one hand it can mean "having no greater extent, range, value or power"; on the other it is what it is in the fullest sense of the term, i.e, "nothing short of." It is in the latter sense that we would wish to approach the notion of the model to try and see it in its plentitude. No doubt this issue and the future issues of this journal will contain many distinguished analyses of particular situations in the form of mathematical models of varying validity and sophistication. Nevertheless we feel that it is worthwhile to lay out some of the philosophical considerations that underlie the whole business of mathematical modelling. It is not that the mathematical modeller needs to be a full-time philosopher, or should attend to the process of forming a mathematical model in the sense of being consciously aware of the philosophic issues involved. It is rather that all who practice the craft of mathematical modelling can benefit by a certain reflection on the underlying notions of the subject. Without such reflections it is indeed possible to get into a measure of confusion about the modelling process. We do not claim to have the last word on the notion of models and modelling, but rather would put forward the following contentions in all modesty as suggestive topics which will bear further discussion and elucidation.

1. THE RISE OF MATHEMATICAL MODELLING

The extent to which mathematical modelling has become self-conscious is in itself a phenomenon of interest. Like Monsieur Jourdain, mathematicians have been modelling for centuries without perhaps recognizing it as their prose. The last few years have seen two international symposia [2], the publication of a number of different books [3–8, to give only a sample] and the founding of journals specifically devoted to mathematical modelling [9–10]. Two encouraging features seem to mark this development. One is the transdisciplinary nature of the enterprise and the other the recognition of the limitations

---

Dr. Aris is with the Department of Chemical Engineering and Materials Science; Prof. Penn is with the University College and Department of Anthropology.

of the process by its own practitioners. It thus may be hoped to conduce to the greater affinity and mutual understanding of different departments of learning, and it is to be hoped also that its progress will not be hampered by the resistance which exaggerated claims inevitably produce.

Evidence of this appears constantly. For example, a symposium on "Mathematical Modelling and its Relevance to the Teaching of Mathematics in Higher Education" was sponsored by the Institute of Mathematics and its Applications at the beginning of 1979. The papers published in the Bulletin of the I.M.A. are of interest in themselves. In the first of these, on "Learning to Use Mathematics" by H. Burkhardt [11], a framework is made for discussing the teaching of mathematical modelling. This is described as "tackling realistic problems from outside of mathematics." The kind of problem that arises in this context is distinguished from the traditional mathematical example by the emphasis that it places on the formulation and on the objective. The objective is not just an exercise in mathematics as learnt but is the solution or understanding of some prototypical problem. There is an analogy here perhaps with the teaching of languages where in the old style short exercises were made up with sentences which would test the students' grasp of the immediately preceeding grammar or syntax. In contrast to this some recent texts have endeavored to use the literature of the language itself to instruct the student in the syntactical and grammatical complexities of the language. R. R. McLone [12] called attention to the report on the "Training of Mathematicians" [1973] in his discussion of the teaching of mathematical modelling. This report emphasized that mathematicians practicing their craft in industry saw a greater role for "mathematical modelling" in their work than had been communicated to them in their education. He points out that all seem to agree that mathematical modelling is more of the nature of an activity than a body of knowledge to be mastered, and goes on to enumerate the skills that are required under the headings of "manipulative," "discovery," "critical," and "communicative." Under the first he recognizes the need for the acquisition of basic techniques, the standard use of these and the extrapolation of them into unfamiliar situations. Under the second he notes the need for improvisation of new techniques to meet a particular situation, the abstraction that often brings together in a unified way diverse problems and the very formulation of problems themselves. Under the heading of "critical" he calls attention to the need for organization of material from standard sources, the assessment of suitability of different methods and the need to be critical of one's own work as well as its sources. The communicative skills include the communication of ideas to mathematicians and nonmathematicians and the effective working together in groups. Indeed the instruction which he favors is of the nature of a mathematical laboratory course. D. Saunders [12] called attention to the course on mathematical modelling which is included in the Open University offerings. On a more technical vein the same issue contains papers by J. G. Andrews on "Mathematical Models in Welding" [14]; by D. Burghes on "Mathematical Modelling in Geography and Planning" [15]; and by D. M. Burley on "Mathematical Modelling in Biology and Medicine" [16].

What is particularly of interest here is the way in which the mathematical curriculum is being affected. It is not that this emphasis on modelling is entirely new—Ben Noble's book on applications of undergraduate mathematics in engineering was published in 1967 [17] and had been preceeded by several years of concern by a committee of the American Mathematical Society—but its time certainly seems to have arrived. It is not that it has not been practiced in disciplines as different as biology [18–23] and geography [24–26], but that modelling is becoming more self-conscious and the validity of its approach being more widely realized. The literature is already vast and obviously cannot be surveyed here; among the chief benefits of the two International Symposia are the impressions they give of the scope of mathematical modelling.

2. SOME USES OF THE WORD "MODEL"

The term "model" has been subject to what Bunge calls a "merry confusion of senses" and, though he has done as much as anyone to clear up this confusion, there are still some ambiguities in the way in which the word is commonly used. Bunge [27,28] recognized that there is a "model object" namely a hypothetical sketch of supposed phenomena, which is to be grafted on to a theoretical model in the process of model making. Thus in confronting the simple case of planetary motion, the model object might consist of two point masses representing the sum and its satellite. If this is grafted on to the Newtonian theory of gravitation there results a Newtonian model of planetary motion. On the other hand, if it is treated from the viewpoint of Einstein's Theory of Relativity, a relativistic model of a simple planetary system evolves.

In his paper in the volume edited by Freudenthal, Apostol [29] formalizes the modelling relationship as $R[S, P, M, T]$ which is the relationship in which the subject S takes, with a certain purpose P in mind, the entity M as a model for the prototype T. This is interesting because it recognizes the role of the modeler, which Apostol calls the "subject," and also brings in the note of purpose. Aris [5] has emphasized that the model must be judged in the light of the purpose for which it was constructed and has also given some attention to the skill of the modeller. It is important to recognize the personal element in the knowledge that might be gained by modelling and we shall return to this point later.

The term "model" is indeed widely used by philosophers of science. Hesse [30] in her excellent little monograph "Models and Analogies in Science" distinguishes two basic senses and Leatherdale, in a longer work [31], four. McMullin [32] has claimed that the "most important single question" posed by physics for philosophy is that of deciding what "the complex postulated structures of the scientist tell us of the world." If this claim can be made for large deliberations on scientific theory, and we believe it can, then perhaps the more particular and detailed considerations that arise in the mathematical modelling of a specific situation can bring something to the greater question—the taxonomist still needs his microscope. For the range of the word "model" must be recognized. McMullin can speak of the theory being "about this model and about nothing else" using "model" in the broad sense, as when mitosis is held to be a matter of exponential growth at a rate conditioned by the availability of nutrients. In the narrower sense we have a "model" when that theoretical vision is focussed on a particular situation, say of competitive growth in a chemostat. The distinction needs to be recognized, but we hesitate to impose it by any stipulation of usage, even for the purposes of this paper, lest we add to the profusion—or, "confusion" as Bunge has it—of current uses.

Speaking loosely, the mathematical model will be the mathematical structure that ties the specific situation back into a more general theory (with its associated physical model). Its validity as an "explanation" of what is going on in the given situation rests on the tripod of: (a) the adequacy of its representation of the situation; (b) its internal correctness; and (c) the acceptability of the general theory which is involved. Thus a mathematical model of a tsunami would be unconvincing if it ignored inertial terms, incorrectly solved the equations or related itself to astrology rather than continuum mechanics. On the other hand, it might with impunity ignore the salinity of the ocean (except in so far as it affects the density) or the relation between continuum mechanics and molecular theory.

3. THE PROCESS OF MATHEMATICAL MODELLING

Let us turn to a brief and informal review of the modelling process. It must be preceeded by some encounter of the subject or modeller with the phenomenon. This encounter will be already somewhat mediated by the theoretical viewpoint of the subject. At this state this may be very general as for example when a person approaches a new phenomenon as something to be explained within the framework of natural science as opposed to taking it to be an act of magic. Nothing very much more than a presumption of, say, the uniformity of nature may be involved here. Let us suppose however, that the modeller in reflecting on the phenomenon conceives a purpose. This might involve an application and require constructing a model which would be of value in predicting its future behavior. Or, the purpose might simply be to understand the possible mechanisms for the phenomenon and tie these back into other known regions of scientific knowledge. Having conceived this purpose, the subject returns to the phenomenon with a more explicit theoretical viewpoint. It is in keeping with the common use of the term to refer to this viewpoint as a "theory." Thus, for example, the subject might approach the phenomenon from the viewpoint of the Newtonian theory of gravitation. This does not preclude subsequent modifications on some other theoretical basis for, if the Newtonian theory failed to give an adequate model, the subject might, as mentioned above, return and approach from the viewpoint of the theory of relativity instead. Our point is that at this stage there is some commitment to a theoretical framework within which the modelling process is to go on. This would seem to correspond to the *theoretical model* of Bunge.

The modelling comes about when some particular assumptions are made about the phenomenon and appropriate methods of description, corresponding to the formulation of the *model object* in Bunge's terminology. These assumptions are essentially of a constitutive nature either concerning the material of the system or the nature of the model which is deemed adequate for the purpose. Again there is a strong personal element in this, for the skill of a mathematical modeller will be seen in the ease and finesse with which these assumptions are made and the way in which an appropriate model evolves. Again, there is plenty of room for iteration and for returning to the situation with a slightly modified set of assumptions. Nor should we overlook the fact that there may be several sets of assumptions which lead to the same model, since in general there is an equivalence class of assumptions which connects the phenomenon, viewed in the light of a general theory, with the mathematical model.

This is not unrelated to the observation that any prototype has a subset of characteristics that are irrelevant to the modelling process, as the ocean's salinity might be to the hydrodynamics of a tsunami. These adiaphora form, as it were, the kernel of the modelling operation which maps the "space" of the prototype onto the "space" of models since they are mapped on the null element and so do not appear.

The mathematical model M is now in existence and from this point has a life of its own. The use that may be made of this intrinsic vitality will depend on the purpose for which the model was intended and what modifications of that purpose may take place. In the simplest case the model may be operated upon to give numbers; as when a solution to a differential equation is found, these numbers can be compared with experimental observations of the phenomenon itself. The model may be judged adequate, if the agreement is within experimental error, or inadequate, if it lies outside of a reasonable estimate of error. An inadequate model may be reformulated by changing the assumptions under which it was constructed. For example, in the modelling of tidal action, it might be found that the results from a model incorporating only the influence of the moon would be inadequate and a modification of the model would be necessary in which allowance would be made for the influence of the sun as well. In other cases the manipulation of the model might be minimal. An example of this would be the rendering of the equations dimensionless so that the important dimensionless numbers (e.g., the Reynolds number from the Navier-Stokes equations) emerge. Another possibility is that the model would suggest a totally new purpose for which further manipulation or modification might be necessary. For example, if an adequate model of tidal action had been obtained for a smooth coastline one might return to this and ask what the effect of promontories and inlets might be. Here again the overarching theory is the same but the hypotheses have to be changed. In Bunge's nomenclature the theoretical model stays the same but the model objects vary.

One might ask whether it is possible to construct models without an overarching theory. The answer would seem to be that this is possible but that the models would remain isolated and the general theory would have to be created anew for each individual model. Thus the general theory serves as a binding element in the space of models.

4. MANIPULATIONS WITH THE MODEL

Once the model is formulated it is open to manipulation which may, as we have just noticed, be as slight as a rearrangement or as serious as a total reshaping. Even when the next objective is the straightforward solution of certain equations there may be need for skill in the approach as Hammersley [33,34] has shown; the preparation of the model and the importance of getting a preliminary feel for the solution has also been emphasized by Aris [5,35,36].

The internal processes of model reduction may be very drastic yet not change the model in essence. Thus a partial differential equation may be reduced to a set of ordinary differential equations as a means of finding the numerical values of the solution. This calls for the technical skills of the numerical analyst, but can hardly be thought to have changed the model. If an analytical solution had been possible it would have been used and would, if the discrete approximation had been done properly, have led to the same numbers to within the tolerated accuracy.

In other cases the transformation of a model is accepted, or even cultivated, for the light it may shed on the situation, for the value inherent in having an alternative approach or even for the sheer aesthetics of variation. Thus in the so-called catalytic monolith a large number of fine passages of identical cross-section pass through a block of heat conducting material so that they form a system of parallel channels separated by plane walls. For simplicity the cross-sections of these channels can be thought of as identical squares. If a hot fluid flows into a central passage and cold into all the rest, the hot stream will lose its heat to the surrounding streams and the temperatures at the other end of the monolith will show a distribution, the central stream still being the hottest with the other temperatures decreasing according to their distance from the central passage. In fact a formula for $T_{m,n}(z)$, the temperature in the (m,n)th passage at a distance z from the inlet, can be derived. If $T_{o,o}(o) = 1$ and $T_{m,n}(o) = 0$, $m,n \neq 0$, we have a "fundamental" solution

$$T_{m,n}(z) = \frac{1}{a^2} e^{-4rz} I_m(2rz) I_n(2rz)$$

where

a = the side of the square

$r = K/V\rho c_p a^2$

K = a transmission coefficient depending on the heat transfer to and through the wall.

On the other hand the fundamental solution for heat conduction in a continuous flowing medium governed by the equation

$$V\rho c_p \frac{\partial T}{\partial z} = k\left[\frac{\partial^2 T}{\partial x^2} + \frac{\partial^2 T}{\partial y^2}\right]$$

is known to be

$$T(x,y,z) = \frac{V\rho c_p}{4} \exp - \frac{V\rho c_p(x^2 + y^2)}{4kz},$$

where

V = velocity of fluid

ρc_p = heat capacity of fluid.

What is the relation between these models, the first a rather literal one, the second metaphoric in the sense of treating a large number of thin walls and narrow passages as if it were a continuum? If a is the length side of square defining each passage we would expect that $T_{m,n}(z)$ would approach $T(ma,na,z)$ as a gets small and this can be shown [37] to be the case provided $k = K$. An equivalent continuum conductivity has thus been calculated which relates the two distinct models. However, the same asymptotic formulae which is used to establish the connection with a different model (the continuous one) could be used merely as computational tools within the original (discrete) model to evaluate $T_{m,n}(z)$.

In this example the reductions involved are highly conventional ones with all the dignity of precise asymptotic formulae. Often, however, the situation calls for approaches that are much more risqué for which the justification is much less certain. Thus Gilles [38], for example, exploits the observed stability of form of a temperature wave in a packed bed, introduces an artificial, though suitable, set of formulae for its shape that allow control calculations that would not otherwise be possible. In these extreme cases it is probably better to regard the model as having been changed by the manipulation into a new model which could be presented in its own right on the basis of its own assumptions.

The skillful use of internal manipulation and model transformation is clearly of the essence of the craft of modelling. But before expanding on this theme let us review what is involved in tying back the model into the prototype.

5. CONFIRMATION OF THE MODEL

Though we have emphasized that the model is an entity in itself and, if nontrivial, deserves study for its own sake, it is often necessary and desirable to relate it in detail to the prototype. The most common way of doing this is to calculate some feature of the behavior of the model which is thought to be representative of an experimental measurement made on the prototype. Thus, for example, one might calculate the resonant frequencies or the critical buckling loads and compare them with experiments. If the observations and the calculation agree to within a reasonable error, then one has a certain degree of confidence that the model indeed "represents" the system. But there are problems with the notion of confirming behavior by means of observation.

The idea of a model is inherent in certain aspects of evaluating everyday evidence. Take, for example, the situation Russell uses to call attention to the difficulty of judging causation by experiment. In two factories widely separated a bell is sounded promptly at noon and the workers take a lunch break. From this observation alone it is not clear that the bell at A is not the cause of the lunch break at B and vice versa. Indeed one can imagine a situation where through lack of facilities at B, the bell is rung by someone telephoning from A to B a moment or two before 12 o'clock and allowing the bell operator to listen to the ringing of the bell at A. In the event of the simultaneous breakdown of the bell at B and in order to work through the lunch hour to fill the rush order at A, the lunch hour at B might be taken by an almost automatic reaction and the observer would see the ringing of the bell at A "causing" the egress of the workers at B. It is the absence of any model which allows such an argument to be carried so far. In common sense terms we would dismiss any causative effect at B of the ringing of the bell at A because the workers there cannot hear it. In other words we invoke a model of the transmission of the signal which permits it only a certain range and recognized that B lies outside of that range. Thus the mere coincidence (even over a long series of observations) of the lunch in one place and the ringing of the bell in the other does not convince us that there is any relation between the two. A model may therefore be part of the explanation even when it is there only implicitly.

What common sense demands of a model which really "fits" the prototype is that there shall be as many connections as possible that tie it back into the prototypical situation. The degree of confidence in the model, or more correctly, in the applicability of the model, grows in proportion to the growth of detail in this connection. An example may serve to illustrate this. In the theory of diffusion and reaction the effectiveness of a catalyst particle is defined as the ratio of the actual reaction rate to the reaction rate which would obtain if there were no limitation by diffusion. Thus the effectiveness is one when there is no diffusion limitation or when the ratio of the reaction rate to the diffusion coefficient is very small. Because the reaction is taking place in the interior of the catalyst pellet, into which the reactant must diffuse, the reaction may deplete the concentration and hence slow the reaction down. If the rate of the reaction is large then the flux into the pellet becomes the limiting factor, and there is a rapid falling off of the concentration as one goes into the catalyst particle. Since the reaction rate is proportional to that concentration, the total reactivity of the pellet is decreased by the diffusive resistance. The effectiveness of the catalyst decreases as the reaction rate increases relative to the diffusion rate and ultimately it becomes inversely proportional to the so-called Thiele Modulus. This modulus is the form $L(K/D)^{\frac{1}{2}}$, where L is the size of the pellet, K is the reaction rate constant and D the diffusion coefficient. Thus the effectiveness, ϵ, is near one when $\phi = L(K/D)^{\frac{1}{2}}$ is small and ϵ is proportional to $1/\phi$ when ϕ is large. Now the total reaction rate per unit volume is proportional to the product of K and ϵ, which, when ϵ is one, is simply proportional to K but, when ϕ is large, the total reaction rate is proportional to the square root of K. K has the form $e^{-E/RT}$ where E is an activation energy, R is the gas constant, and T is the absolute temperature. Thus for small values of ϕ the reaction rate is proportional to $e^{-E/RT}$ whereas for large values of ϕ it is proportional to $e^{-E/2RT}$. It follows that if the logarithm of the reaction rate is plotted against the reciprocal temperature, the curve would have a slope of -1, for large values of the reciprocal, whereas for small values of the reciprocal temperature it would have a slope of $-\frac{1}{2}$ (N.B. the temperature is large when its reciprocal is small and vice versa).

Now the single observation that, over a sufficiently wide range of temperature, the dependence of the reaction rate on the reciprocal temperature had this transition from slope $-\frac{1}{2}$ to slope -1 would give one a first degree of confidence in the model of the catalyst particle on which this computation is based. Such in fact; was the confirmation that established the validity of the model in its early days [30]. Indeed one might go further and deduce the ratio of certain physical quantities by fitting the data by a pair of straight lines of slopes $-\frac{1}{2}$ and -1. However, a much more important confirmation of the model would arise if measurements of the diffusion coefficient of the reactant in the porous material were made. Alternatively, this diffusion coefficient could be calculated and measurements made of the intrinsic rate constant. Then, from the formula for the effectiveness that the model gives, the reaction rate could be calculated and compared with observation. If these two agree to within the experimental error then one may have a much greater degree of confidence in the model.

This example illustrates the fact that the degree of confidence in the applicability of a model is greatly raised when there is a close correspondence with relevant observations and the estimation of the parameters in the model has been done independently. If the correspondence of the model and observation is used to determine the parameters then there is a lesser degree of confirmation of the model for very much less independent information has been supplied. Some degree of confirmation still exists, for a totally unsuitable model might give no correspondence in such a test as the rate vs reciprocal temperature plot for the catalyst particle.

Now let us suppose that some particular dynamic behavior has been predicted by the model which, it is hoped, can be related to experiment. If the initial conditions of the experiment and of the calculated behavior are close one might hope that the subsequent behavior of the two would remain close. Such is the case in many instances of stable motion, but even a periodic response presents some difficulties. How would one confirm a periodic response? First of all it would be necessary, in comparing the ultimate response to a forced oscillation to allow any transients in the system to die out, for these would be peculiar to the starting conditions. Secondly, one would ignore the phase of the resulting oscillations, since this would be dependent on the particularities of the start of the process. Then one would attempt to match up the wave form, its period and particular characteristics. But, even assuming very accurate observations, there are some problems with wave forms of sufficiently long period.

The problem becomes even more acute when one considers so-called chaotic motion. "Chaos" in a dynamical system is sometimes defined as a behavior which is neither steady, nor periodic, nor asymptotically approaching a periodic behavior. It has received much attention from mathematicians and engineers since the discovery of a simple system of three ordinary differential equations by Lorentz [40]. It has an appearance of randomness which is not of a statistical nature. For in a nonchaotic system subject to certain random fluctuations, two trajectories starting close to one another would retain some resemblance when allowance was made for the fluctuations due to the random element. By contrast, two trajectories starting close together in a chaotic system would, in general, bear no resemblance to one another after a sufficient time. The difficulty in comparing a prediction with observation lies in the fact that chaotic solutions often coexist with other solutions of very long periods or with almost periodic solutions. Thus given a long and apparently chaotic record it is impossible to assert that this is not part of an even longer periodic solution. Here the role of the model would seem to be crucial in having any confidence in the experimental result, or, for that matter, in the model. One would need to tie the model to the prototype in as many ways as possible and to show that certain ranges of the parameters in the model gave rise to the same kind of irregular behavior as could be observed in the real system for more parameter ranges.

However, there is an even graver philosophical problem which may be illustrated from some results in the theory of oscillators. The so-called Van der Pol equation gives rise to oscillations either when left to itself or when excited with a simple harmonic or periodic impressed excitation. The Van de Pol equation is of the form $\ddot{x} + f(x)\dot{x} + x = bp(t)$. If $b = 0$, its solution is a relaxation oscillation, but with excitation has a very complex response for certain ranges of the amplitude, b. In the space of x, $y = \dot{x}$ and t, the time, the solution is represented by a point on a line always proceeding with unit speed in the t direction. A return map can be made in which any given point x, y in a plane such as $t = 0$ is mapped into a corresponding point in the x, y plane at $t = 2\pi$. This return map shows the difference of state at the beginning and end of one period of the excitation and is a mapping of the x, y plane into itself. Now it has been shown that the following property obtains for certain solutions of this equation. In the x, y plane let us construct two regions, small rectangles about two points which will be referred to as A and B. In the wandering of the point given by the return map we write down A if the point enters the box A and B if, after a sufficient number of periods it enters the box B. This corresponds to noting that the state of the system comes close to some observation point corresponding either to the box A or the box B. The size of the box might be a measure of the accuracy of possible observation. The motion can be chaotic in the following sense. Given any sequence of the letters A and B, such as $AABBBABAAABA$, it is always possible to find a point within the rectangle A such that the sequence of entries into either the box A or the box B is as prescribed. What is more, this is true if time is reversed and the trajectory is run backwards. It follows that the future is in some sense independent of the past for both can be prescribed and then "observed." Of course there will be many other trajectories which have different sequences of entry into the two boxes but, assuming sufficient powers of observation one can always look for and find any desired result in this system. One could almost say that this was a confirmation of the model, namely that one could eventually find any required behavior, save for the fact that similar results might be shown by a totally different system of equations.

In turning from physical to social prototypes an interesting distinction arises from the very existence of the prototype's self-consciousness. In this case it may have a model of itself and we have to ask what obligation the modeller has to this prior model. For example, in modelling the kinship structure of a society must an anthropologist's model though no doubt couched in different terms be faithful to the model that the society has of its own structure? Is this fidelity, even when appropriate, of a strict nature, as when $g(x)$ is a restriction of $f(x)$? Or can it be much looser, as when the society has a mechanical model of itself but the anthropologist builds a statistical model? In one of the places where Levi-Strauss draws this distinction [41] he also points out the dual nature of structural studies, or as we might say, interrelationships of models. On the one hand, they isolate certain phenomena and bring special methods to bear on them. On the other, they seek to construct abstract models which can be put into some sort of isomorphism with other models. This isomorphism (at least to the extent that it carries conviction) is held to have a certain explanatory power [41, p. 277]. In physical sciences the explanatory status of the mechanical analogues in electromagnetic theory—beloved of the 19th century English physicists which Duhem found so derisory—is equally debatable, though the isomorphism can usually be drawn more convincingly. Perhaps it is having a self-conscious subject of study that most clearly differentiates the social scientist from the natural; so that, however much the one may learn from or use the methods of the other, the "wirtschaftler" is engaged in a different kind of reflection from the "wissenschaftler." The economist and statistical mechanician may lie down together, but the anthropologist will not eat straw like the graph theorist; it is to be hoped that none will hurt and destroy—at least not too much.

6. MODELLING AS CRAFTSMANSHIP

The emphasis we have put on mathematical modelling as a craft demands a brief, further discussion. We believe that it deserves consideration, though what we have to say here is preliminary in the strictest sense of the word. What goes on in the modeller's head is not purely formalizable, either in the more abstract terms of Apostol [29] or Bunge [27] or in the taxonomic vein of Harré [42]. Nor is it purely subjective, conditioned by sociology and psychology as Lakatos [43] and Feyerabend [44] would have us believe. It has more in common with the tacit element in personal knowledge that Polanyi [45,46] has drawn out. It has structure, it has techniques that can be taught and learned, but involves also a personal touch, not only in trivialities (such as the choice of notation) but in deeper considerations of skill and suitability. It also involves an element of risk, since a wrong turn in the development of a model may lead to its complete stultification. Like the furniture maker, the mathematical modeller shapes the several parts of his work and fits them together. If one is marred or misshapen it must be reworked or even discarded and replaced.

Pye [47,p.4] has brought out this aspect of craftsmanship very well in his book, *The Nature & Art of Workmanship*. "If," he writes, "I must ascribe a meaning to the word craftsmanship, I shall say a first approximation that it means simply workmanship using any kind of technique or apparatus, in which the quality of the results is not predetermined, but depends on the judgement, dexterity and care which the maker exercises as he works. The essential idea is that the quality of result is continually at risk during the process of making; and so I shall call this kind of workmanship 'The workmanship of risk'; an uncooth phrase, but at least descriptive." He goes on to contrast this with the workmanship of certainty in which the quality of the product is largely predetermined. The latter is exhibited by machine printing in contrast to handwriting, for the control of quality is reduced to a few settings once the compositor's is locked up in the frame.

The contrast has an interesting parallel in Zeigler's description of "modelling" and "simulation" [48]. The former relates the so-called real system to the model and vice versa; the latter is the simulation of the model on the computer. The modelling, like handwriting, is the workmanship of risk; the simulation, like machine printing, that of certainty. If it is argued that the simulation involves the craft of computer programming then we may reply that printing involves the compositor's craft which, like the programmer's, has elements both of risk and certainty. But once a program is written and running the quality of outcome is predetermined, though not prejudged—you may have an inaccurate set of numbers as easily as a badly printed newspaper.

Many a paper and book on mathematical modelling [3,4,5,7,11,12,13,48] has its diagram of the modelling process in which the relation of the model to the "real" world and the iterative nature of the process are portrayed. We have resisted the urge to devise yet another, though the temptation to make a culinary triangle [49, see p. 29] out of Bunge, Feyerabend and Polanyi (or rather, their positions) is strong. Rather, some of the ideas nascent in the craft of mathematical modelling have been discussed in the hope that they may contribute something to the philosophers' ongoing discussion of the concept of a model [27–32,41–44,50].

Between the instrumentalist's position (that models serve only as computational tools) and the realist's (that models actually tell us something about the real world) there is the need for a via media that will do justice to the prehensions of the modeller and to the judgement, skill and purposefulness that are tacit in a well-constructed model. Nor is it only a matter of what goes on in the modeller's own head, for modelling is often a cooperative and even a transdisciplinary affair. Indeed it may well be that this dialogue is the appropriate model for the individual modeller's internal prehensions.

In his *Mathematician's Apology,* G. H. Hardy claimed victory over A. E. Houseman in the matter of priority the latter had ascribed to the critic. To Hardy, as to lesser mathematicians of a more applied cast of thought, the creative act, however humble, takes precedence over philosophical commentary, however profound. So let it be here. As the pages of this journal unfold we shall see the mathematical models aplenty in all their variety and vigor. *They* will make the best contribution to the evolving concept of the mathematical model.

Acknowledgement—The authors are indebted to Professor S. Gudeman, Department of Anthropology, University of Minnesota, for some helpful discussions.

REFERENCES

1. *The Oxford English Dictionary,* Vol. 6, pp. 353–4 (senses 4 and 5). Clarendon Press, Oxford (1933).
2. X. J. R. Avula, ed., *Proceedings of the 1st and 2nd International Symposium on Math Modelling*. St. Louis, Missouri (1977, 1979).
3. J. G. Andrews and R. R. McLone, *Mathematical Modelling*. Butterworth, London (1976).
4. R. Haberman, *Mathematical Models*. Prentice-Hall, Englewood Cliffs, N.J. (1977).
5. R. Aris, *Mathematical Modelling Techniques*. Pitman, London (1978).
6. M. J. Lighthill, ed., *Newer Uses of Mathematics*. Penguin, New York (1978).
7. E. A. Bender, *An Introduction to Mathematical Modeling*. John Wiley, New York (1978).
8. H. Burkhardt, ed., *The Real World of Mathematics*. Shell Center for Mathematical Education, Nottingham University, Nottingham (1979).
9. C. A. Brebbia and J. J. Connor, eds., *Applied Mathematical Modeling*. IPC Science and Technology Press, Guildford, England (1977).
10. D. N. Burghes and G. M. Read, eds., *Journal of Mathematical Modelling for Teachers*. Mathematics Dept., Cranfield Institute of Technology.
11. H. Burkhardt, Learning to use mathematics. *Bull. of the Inst. of Math. and Applics* **15,** 238 (1979).
12. R. R. McLone, Teaching mathematical modelling. *Bull. of the Inst. of Math. and Applics* **15,** 244 (1979).
13. D. Saunders, Mathematical modelling at the Open University. *Bull. of the Inst. of Math. and Applics.* **15,** 247 (1979).
14. J. G. Andrews, Mathematical models in welding. *Bull. of the Inst. of Math. and Applics.* **15,** 250 (1979).
15. D. Burghes, Mathematical modelling in geography and planning. *Bull. of the Inst. of Math. and Applics.* **15,** 254 (1979).
16. D. M. Burley, Mathematical modelling in biology and medicine. *Bull. of the Inst. of Math. and Applics.* **15,** 261 (1979).
17. B. Noble, *Applications of Undergraduate Mathematics in Engineering*. Macmillan, New York (1967).
18. J. M. Smith, *Mathematical Ideas in Biology*. Cambridge University Press, Cambridge (1971).
19. J. M. Smith, *Models in Ecology*. Cambridge University Press, Cambridge (1974).
20. N. Rashevsky, *Mathematical Biophysics*. Dover, New York (1960).
21. H. H. Lieberstein, *Mathematical Physiology*. Elsevier, New York (1973).
22. F. Heinmets, ed., *Concepts and Models in Biomathematics*. Marcel Dekker, New York (1969).
23. J. D. Murray, *Lectures in Nonlinear Differential Equation Models in Biology*. Oxford University Press, Oxford (1977).
24. R. J. Chorley and P. Haggett, eds., *Models in Geography*. Methuen, London (1967).
25. B. F. Fitzgerald, *Development in Geographical Method*. Oxford University Press, Oxford (1974).
26. A. G. Wilson and M. J. Kirkby, *Mathematics for Geographers and Planners*. Clarendon Press, Oxford (1975).
27. M. Bunge, *Method, Model and Matter*. Riedel, Dordrecht (1973).
28. M. Bunge, *Treatise on Basic Philosophy*. Vols. 3 and 4, *Ontology,* Riedel, Dordrecht (1977, 1979).
29. H. Freudenthal, ed., *The Concept and Role of the Model in Mathematics and the Social Sciences,* Riedel, Dordrecht (1961).
30. M. B. Hesse, *Models and Analogies in Science*. Sheed and Ward, London (1963).
31. W. H. Leatherdale, *The Role of Analogy, Model and Metaphor in Science*. North-Holland, Amsterdam (1974).
32. E. McMullin, What do physical models tell us? In *Logic, Methodology and Philosophy of Science*. North-Holland, Amsterdam (1968). pp. 385–396.
33. J. M. Hammersley, *Maxims for manipulators. Bull. of the Inst. of Math. and Applics.* **9,** 276 (1973); **10,** 368 (1974).
34. M. M. Hammersley, How is research done? *Bull. of the Inst. of Math. and Applics.* **10,** 214 (1974).
35. R. Aris, How to get the most out of an equation without really trying. *Chemical Engineering Education,* **10,** 114 (1976).
36. R. Aris, *Re, k* and π: A conversation on some aspects of mathematics modelling. *App. Math. Modelling* **1,** 386 (1977).
37. R. Aris, De exemplo simulacrorum continuorum discretalumque. *Arch. Rat. Mech. Anal.* **70,** 203 (1980).
38. D. Gilles, Model reduction. In *Dynamics and Modelling Reactive Systems*. Academic Press, New York (1980).
39. R. Aris, *The Mathematical Theory of Diffusion and Reaction in Permeable Catalyst*. Clarendon Press, Oxford (1977).
40. E. N. Lorentz, The problem of deducing the climate from governing equations. *Tellus* **16,** 1 (1964).
41. C. Levi-Strauss, *Structural Anthropology*. Doubleday, New York (1967).

42. R. Harré. *The Principles of Scientific Thinking*. University of Chicago Press, Chicago (1970).
43. I. Lakatos and A. Musgrave, *Criticism and the Growth of Knowledge*. Cambridge University Press, Cambridge (1970).
44. P. Feyerabend, Consolations of the specialist. In *Criticism and the Growth of Knowledge*. Cambridge University Press, Cambridge (1970), pp. 197–230.
45. M. Polanyi, *Personal Knowledge*. Routledge, Kegan, Paul, London (1962).
46. M. Polanyi and H. Prosch, *Meaning*. University of Chicago Press, Chicago (1975).
47. D. Pye, *The Nature and Art of Workmanship*. Cambridge University Press, Cambridge (1968).
48. B. P. Ziegler, *Theory of Modelling and Simulation*. Wiley, New York (1976).
49. E. Leach, *Claude Levi-Strauss*. Penguin, London (1971).
50. G. H. Muller et al. Modellbegriffe, I-V [A series of papers]. *Studium Generale* **18,** Fascs. 3–7 (1965).

Ut Simulacrum, Poesis

Rutherford Aris

> Etenim omnes artes, quae ad humanitatem pertinent, habent quoddam commune vinculum et quasi cognatione quadam inter se continentur.[1]

If Cicero's somewhat orotund statement in defense of Archias is to be taken seriously it leaves us in a quandary. Either it is an assertion so general that it may be taken as axiomatic, or it is a proposition so comprehensive that it would take volumes to justify it. Nevertheless it is a viewpoint that comes naturally within a university, where, if anywhere, there should live on a sense of the unity of knowledge and of the juice and joy of its continual discovery and refreshment. To develop any such large picture is clearly beyond the scope of this essay, but I would like to try to outline some "common bonds" which I believe exist between the activities of the poet and those of the mathematical modeler, for it is such *simulacra* that I have in mind in the title. Like William Jackson, I "would keep the principle within its proper bounds,"[2] for it is not without ample justification that Hagstrum has referred to Horace's phrase "ut pictura poesis" as a text too often used as a pretext "for riding critical hobbyhorses."[3] And if this be done in the green tree of the accurate quotation, what shall be done in the dry branch of its deliberate modification?

Let me say at the outset that I want to avoid both grand theory and detailed analogy—the one because I am no philosopher and have not the nous for it, the other because I believe it would trivialize the subject and lead inevitably to a ridiculous final position. Rather than a pointwise analogy in which an attempt is made to tie details together, I would emphasise the *fellowship* (in the most literal sense of the word) that exists between *makers,* whether of poems or of models, somewhat in the spirit of Sackville-West's

> All craftsmen share a knowledge,
> They have held reality down fluttering to a bench;
> Cut wood to their own purposes; compelled

> The growth of pattern to the patient shuttle;
> Drained acres to a trench.[4]

After a brief look at some of the comparisons that have been made under the *pictura* rubric, I shall turn to mathematical modeling and introduce a population model which I hope will be found sufficiently elementary to be easily comprehensible and sufficiently deep to sustain the weight of my argument. In particular, I want to show that the notions of aptness, intrinsic standard, internal tension, and iterative nature find echoes in poetic craft. The nature of the comparisons I wish to draw is perhaps best expressed as the kindred "affections" of the scientist and poet to which Wordsworth drew attention in the 1802 version of his Preface to *Lyrical Ballads.* He found the knowledge of the one remote and of the other familiar, but claimed that "we have no knowledge, that is, no general principles drawn from the contemplation of particular facts, but what has been built up by pleasure, and exists in us by pleasure alone." Poetry by its very familiarity "is the breath and finer spirit of all knowledge: it is the impassioned expression which is in the countenance of all Science."[5]

The tag from Horace, however, has served as motto for a whole tradition of comparative literary theory that goes back more than twenty-four centuries, for Plutarch attributes to Simonides of Ceos the epigram,[6] already a commonplace, that the writer of *Ad Herennium* quotes as "poema loquens pictura, pictura tacitum poema debet esse."[7] Though earlier in the *Ars Poetica* Horace uses the analogy of painting and poetry in other ways, at this point he is merely stressing that the picture that only does well in the shadows or cannot stand repeated observation is, like the poem that cannot bear the light of criticism or that pleases but once, inferior. "It is innocent, if it is not actually opposed" to any "far-reaching assertions on the relation of the arts,"[8] yet it has been repeatedly used as a vexillum in the advance of diverse critical battalions: by humanists of the Renaissance to emphasize the superiority of painting; by Lessing to put poetry and painting back in their respective domains of time and space; by Babbitt to rehumanize them (at any rate as he saw it).[9]

One remarkably long-lived work in the history of pictorialist theory was the Latin poem of Alphonse Du Fresnoy, *De Arte Graphica,* written in 1637 and published posthumously in 1661,[10] whose popularity Hagstrum regards as an index of the popularity of pictorialism in the seventeenth and eighteenth centuries.[11] Du Fresnoy "thought it improper to publish it without a French translation, which he deferred undertaking from time to time, out of diffidence of his own skill in his native language, which he had in some measure lost by his long residence in Italy."[12] It appeared in England in 1695 translated into prose by John Dryden with a rather labored preface, *A Parallel of Poetry and Painting.*[13] It is not Dryden at his best,[14] for he had "borrowed only two months" from his work of translating Virgil to undertake it at the urging of "many of our most skilful painters and other artists," but Pope thought well enough of it to commend it to his friend and instructor in painting, Charles Jervas, as ". . . instructive leaves, in which conspire / Fresnoy's close Art, and Dryden's native fire."[15] Du Fresnoy combines Horace and Simonides in his first four lines and concurs with a common Renaissance punctuation of Horace which associates the fourth word *erit* with *ut pictura poesis,* instead of the succeeding sentence.[16] This reading, which can be traced back to a fifth century gloss, gives a rather more forceful tone to the analogy than perhaps Horace would allow. Thus Du Fresnoy's

> Ut pictura poesis erit; similisque poesi
> Sit pictura; refert par æmula quæque sororem,
> Alternantque vices et nomina; muta poesis
> Dicitur hæc, pictura loquens solet illa vocari.

becomes in Dryden's prose: "Painting and Poesy are two sisters, which are so alike in all things, that they mutually lend each other both their name and office. One is called dumb poesy, and the other a speaking picture."[17]

In his introduction Dryden draws the cords of analogy fairly tightly, starting with the contention "that one main end of Poetry and Painting is to please" and considering the "rules which may direct them to their common end," for in fidelity to their imitation of nature lies their road to perfection. The principal part of their common art is "Invention," without which the "Painter is but a copier, and a Poet but a plagiary." In neither is this invention unbridled but held in harmony with the "texts of antient authors," rejecting "all trifling ornaments" and avoiding everything "which is not proper or convenient to the subject." The second part is "Design or Drawing," where variety of form and face mirrors the variation of character in poem or play, where absurdities are avoided and passions are congruous. Principal figures are to be central and in the principal light, with groups, like episodes, in support, but these "groupes must not be on a side, that is with their faces and bodies all turned the same way," just as "in a play, some characters must be raised to oppose others, and to set them off the better, according to the old maxim, *Contraria juxta se posita, magis elucescunt.*" The third part of their art is "Chromatic or Colouring," to which "Expression and all that belongs to words" correspond, for does not Horace himself call them "Operum Colores"?

Thus strong colors are like bold metaphors and "Good Heavens! how the plain sense is raised by the beauty of the words," which "a true poet often finds, as I may say, without seeking: but he knows their value when he finds them, and is infinitely pleased." The lights and shadows that belong to "Colouring" put Dryden in mind of a line in Horace that follows "Ut pictura poesis," namely "Hoc amat obscuram, vult hoc sub luce videri,"[18] which he interprets as justification for the contention that the most beautiful parts of picture and poem must be the most finished, whereas there are many things in both which, not deserving of this care, "must be shifted off, content with vulgar expressions."[19] Finally, both Painter and Poet must know when to "give over, and to lay by the pencil."

I have summarized the drift of Dryden's analogy and the style of his argument,[20] not to imitate it, but because I want to try a slightly different approach. Mischa Penn and I have argued elsewhere that the notion of craftsmanship in mathematical modeling provides a via media between the formalistic and subjective extremes, at once doing justice to the modeler's mental activity and providing a basis for discussion.[21] I want therefore to suggest that there is a certain sympathy of intention and compatibility of craft in the work of the poet and of the mathematical modeler. I am not claiming any explanatory value for the comparison, nor that one is a model for the other, but that the poet and modeler should resonate to each other's activities and rejoice in each other's joy, which I take to be the "chief end" of both: "I too will something make / And joy in the making."[22] Each sets out to construct an entity—the one literary, the other mathematical—which will do justice to the maker's vision. The scale may be small, as in the elementary population model I shall describe or Housman's "Epitaph on an Army of Mercenaries," or large, as when a model can rightly be extended into a veritable *Dynasts,* usually under such a title as *The Mathematical Theory of Demography.* Mathematicians, like poets, may be naive or sentimental, grave or gay; their diction may be commonplace or exalted; their metrics as tight as Bridges's or as experimental as Hopkins's. Both, though, would seem to share the common objective of clothing their thought in the dress meet for it. The mathematician responds instinctively to Dylan Thomas's "In my craft or sullen art." He knows that sullen protectiveness of an idea, crouches over it like a diamond cutter over the rough stone, seeking the cleavage plane of truth that will reveal the full beauty of it with the least tap. By taking an abstract point of view, the mathematical modeler, like the poet, is approaching his subject "by indirection."[23]

If this sounds nebulous,[24] I must try to infuse it with some substance by giving a concrete example of a mathematical model which is both elementary and nontrivial. First, however, I should say what a mathematical model is. It is a system of mathematical equations, usually excogitated from some other situation, which has a life of its own by exhibiting a behavior sufficiently rich and appropriately relevant. The "other situation" may be a physical one, some aspect of the non-mathematical world (for example, a linguistic phenomenon) or some other mathematical construct. In this way it is usually "mimetic" in the sense of imitative, but, insofar as it has its own independent existence, it is "mimetic" in Hough's understanding of the Aristotelian sense of being a construct of the imagination.[25] If the purpose of the model is to imitate a natural situation with great fidelity, perhaps to predict its future behavior, then it will be judged by the accuracy of its correspondence;[26] but a model can also serve the purpose of defining, clarifying, and enriching a concept. Though the word *model* might be thought a little demeaning to the larger branches of pure mathematics, there is a continuum of thought that connects the most mimetic model with the farthest reaches of "thatt supersensuous sublimation of thought, / the euristic vision of mathematical trance."[27] G. H. Hardy's fierce denunciation of ballistics and aerodynamics as "repulsively ugly and intolerably dull" was motivated by his pacifist convictions rather than his love of "the 'real' mathematics of 'real' mathematicians,"[28] and the distinction between "pure" and "applied" mathematics is no longer tenable. "Poetical Reason" may not be "the same as mathematical Reason,"[29] but in the motive power of imagination they have a commonality that transcends reason.[30]

The simplest of population models demands only the use of the most elementary algebra since it looks at the population in a certain environment at discrete intervals of time, the successive breeding seasons which we will simply call "years." The model may apply to any population in a fixed region, such as wildebeest in the Serengeti or bacteria in a Petri dish, so long as its growth equation is appropriate, but it also has a life of its own as an example of a so-called difference equation or recurrence relation. We denote by p_n the population at some particular point (say the end) of year n; so p_0 is the population at the start of the period under consideration, p_1 is the population after one year, p_2 after two years, and so on. (A p with no suffix will be used for a typical population when the year in question is not important.) A law of growth is a formula that tells us the relationship between the population in year n and the prior history of the population, that is, p_n is calculable if $p_{n-1}, p_{n-2}, \ldots p_1, p_0$ are known. The simplest case is when this year's population depends only on last year's, or p_n is calculable if we know p_{n-1}. In this case it would be a simple matter to calculate the evolution of the population if we knew what it was to begin with, for from p_0 we could calculate p_1, from p_1 we would get p_2, and so on.[31]

There is a whole family of medieval bestiaries whose description of the elephant begins uncompromisingly "Est animal quod dicitur elephans in quo non est concupiscencia coitus."[32] The elephant, so these bestiaries tell us, is monogamous and has only one life-partner. Indeed, so free is the male from all carnal desire that the female has to seduce him with the mandragora root to secure the one mating of her lifetime, the which accomplished she gives birth in due time to a single offspring. What the successive generations of monks who copied these manuscripts failed to recognize was that the elephant population would be halved at each generation and that the species would rapidly be extinguished—an initial population of a million would last only 19 generations, for example. The formulae for this population history would be $p_1 = p_0/2$, $p_2 = p_1/2 = p_0/4$, $p_3 = p_2/2 = p_0/8$, and in general $p_n = p_0/2^n$. The converse situation is afforded by the legend of the reward claimed as one grain of rice on the first square of the chess board, two on the second, four on the third, and so on, from a monarch who little suspected that the total would be over 18 billion billion grains in all. This might be the case of a population that divided by binary fission in each generation; indeed the word *fission* reminds us that it is the algebra of the atomic bomb. Here we would have a population doubling each time, so $p_1 = 2p_0$, $p_2 = 2p_1 = 2^2p_0$, $p_3 = 2p_2 = 2^3p_0$, or $p_n = 2^np_0$.

Both these are cases of the same law, that of exponential growth or decay. The population at each generation is a constant multiple, a, of the previous population, $p_n = ap_{n-1}$ and so $p_n = a^np_0$. If $a > 1$, the population grows exponentially (indeed, explosively if a is much greater than one) whereas if $a < 1$ it decays (it might be said to be snuffed out if a is very much less than one).[33] The number a is called a parameter of the problem, the "growth parameter," if you will, for decay can be regarded as an inverse growth. It can be thought of as the net difference between a birth rate and a death rate. Suppose that, when the population is p, the number of births is bp. Then, in the absence of death the growth rate would be $a = 1 + b$ since the population p persists and is augmented by the births bp. If the number of deaths, when the population is p, is dp, then the growth $(1 + b)p$ would be diminished by dp giving $a = 1 + b - d$. This is the simplest model of population growth which lies behind the crudest Malthusian fears. Only if the birth rate equals the death rate $(b = d)$ so that a is exactly equal to one $(a = 1)$ will the population remain constant at whatever value it had at the beginning, $p_0 = p_1 = p_2 = \ldots p_n$.

We see that there is no natural population level in the problem, for p_0 can be any number we please. The population level is stable when the birth rate just balances the death rate, for, were we to add to the population by artificially introducing members of the same species, the population would remain at this new level, it would neither explode nor go into a tail spin. Thus the steady population, though arbitrary and stable, is sensitive in the sense that the slightest change in the growth rate parameter a will result either in exponential growth or in exponential decay. This kind of instability can be called parametric sensitivity since the behavior of the population changes with the slightest change in the parameter. If the birth rate exceeds the death rate ($b > d$, $a = 1 + (b - d) > 1$) the population grows without bound, clearly an unstable situation; if the death rate exceeds the birth rate ($d > b$, $a = 1 - (d - b) < 1$) the population declines, a stable enough situation but with the stability of death. The extinct state ($p_n = 0$, for all n) is always a solution of the equation $p_n = ap_{n-1}$, but not a very interesting one; mathematicians call it the "trivial solution." If the death rate exceeds the birth rate ($a < 1$) this extinct steady state is stable, for if we artificially introduced a few of the species they would die out and the population would return to its steady state of extinction. This is a stronger concept of stability than we had with the case $a = 1$. There the population did not return to its value before the artificial disturbance, though it did stay within bounds. Here, with the trivial solution and $a < 1$, the population actually returns to its steady state $p_n = 0$. Mathematicians call this stronger kind of stability "asymptotic stability." By contrast, if the birth rate exceeds the death rate ($a > 1$) the extinct or trivial steady state is unstable. For if we artificially introduce some of the species, the population will grow without bound and never return to the extinct steady state.

To summarize this first stage of the population model, we have a concept of growth (and its inverse, decline) which we have related to birth and death rates and have embodied in a parameter a. If $a < 1$, the only steady state is the extinct state and that is stable, for all perturbations die away into extinction. On the knife-edge balance $a = 1$, when birth and death exactly cancel each other, any population can exist as a steady state, stable in the sense that perturbations do not run away, but not asymptotically stable and sensitive to the least change in the parameter a. If $a > 1$, the only steady state is the trivial one of extinction, but this is unstable and any perturbation leads to unbounded growth.

Clearly this model (to be known as the "linear model," since $p_n = ap_{n-1}$ is a linear equation) has its limitations. Though extinction is a real enough possibility, unbounded growth is not. It is mathematically conceivable and correctly formulated and so a valid model in itself, but it fails to do justice to the situation, it is not condign. We do not value as great poetry the Victorian hymn on Jonah worshiping in the whale's belly which contains the stanza:

> Ah me! This is an awesome place
> Without e'er coal or candle,
> Nothing but fish's tripes to eat
> And fish's tripes to handle.

This is not because it is ill-constructed or entirely unaware of the overtones of its words. "Tripes" is ill-chosen on any reckoning, but "awesome," for example, is used wittingly, for it had not yet been drained of its sense of the numinous to the extent that now allows it to qualify "statistics" or the office of the Vice-Presidency of the United States. Rather it is the inappropriateness of its tone. It might do (with a couple of capitals) for a Law School skit (as the plaint of a student unable to break out of his interpretive community) but scarcely meets the dignity even of the tin chapel. What poetry would be condign? I cannot recall another hymn that takes up this particular story, but it would be hard to compete with the version of 1611, "The waters compassed me about, even to the soul: the depths closed me round about, the weeds were wrapped about my head."[34]

This example is perhaps a little extreme, and a much subtler one is given by Housman in his preface to Manilius V. He first encountered Walter de la Mare's "Fare well" in a newspaper review, printed thus:

> Oh, when this my dust surrenders
> Hand, foot, lip, to dust again,
> May these loved and loving faces
>   Please other men!
> May the rustling harvest hedgerow
> Still the Traveller's Joy entwine,
> And as happy children gather
>   Posies once mine.

"I knew in a moment," writes Housman, "that Mr de la Mare had not written *rustling,* and in another moment I had found the true word. But if the book of poems had perished and the verse survived only in the review, who would have believed me rather than the compositor? The bulk of the reading public would have been perfectly content with *rustling,* nay they would sincerely have preferred it to the epithet which the poet chose. If I had been so ill-advised as to publish my emendation, I should have been told that *rustling* was exquisitely apt and poetical . . . and I should have been recommended to quit my dusty (or musty) books and make a belated acquaintance with the sights and sounds of the English countryside. And the only possible answer would have been *ugh!*"[35] Here are poetical sensibilities of a high order at work.[36]

To develop the model we need to recognize that the growth rate parameter a will be affected by the environment. Most environments will sustain only a finite population and when it is reached the population will, so to speak, choke to death. This can be most simply recognized by putting $a = c(P - p)$, where c is a new parameter, P is the saturation population and p the current population. The quantity $(P - p)$ is the difference between these two populations or the distance from saturation. This goes to zero as the population approaches saturation and so takes the growth rate parameter to zero. For example, if $P = 1000$ and $c = 1/500$, the growth rate parameter a would be close to $2 = 1000/500$ when p is small, say 10–40, but a is less than 1/10 if p is above 950. The equation that governs the evolution of the population is $p_n = c(P - p_{n-1})p_{n-1}$, an equation known as the "logistical equation." It affords a very much richer range of behavior than the linear model.

A very important idea in the development of a mathematical model is that of giving the magnitude of the variables a meaning intrinsic to the problem. If I say $p = 1000$, you do not know whether that is large or small: it may be large for bears on the polar ice and small for bacteria in the laboratory. If however I say $p = 0.95P$, you know immediately that it is a large population, so large in fact that the environment is 95 percent saturated. By the internal standards of the problem this is clearly a large population, whereas $p = 0.05P$ would be a small population. We therefore make the population dimensionless by putting $p = Px$ or $x = p/P$. Thus $p_n = c(P - p_{n-1})p_{n-1}$ becomes $Px_n = c(P - Px_{n-1})Px_{n-1}$ and, when we divide through by P and write $g = cP$, we have

$$x_n = gx_{n-1}(1 - x_{n-1}).$$

Not only does x have a real meaning and an appreciable magnitude, but this manipulation has revealed that there is a characteristic growth parameter g. Since $a = c(P - p)$ was the old growth parameter, we see that the new characteristic growth parameter $g = cP$ is the value of a when p is negligible compared with P, that is, x is virtually zero. Another way of looking at the characteristic growth parameter g is to say that it is the greatest value that the growth parameter a can have, for $c(P - p)$ is greatest when p is least, that is, when $p = 0$.

If dimensionless numbers give magnitudes a meaning within the context of a model, does not the poet do the same sort of thing when he expresses conditions or emotions within the context of the poem? Thus Keats in "The Eve of St. Agnes" would be far less effective had he left the reader to judge the conditions from the plain statement of the first line:

> St. Agnes' Eve—Ah, bitter chill it was!
> The owl, for all his feathers, was a-cold;
> The hare limped trembling through the frozen grass,
> And silent was the flock in wooly fold.

The second line, reading as it does like a shiver, makes the reader feel the cold with an intensity that almost needs the more expectable standards of the next two to moderate its impact. Shakespeare needs not even the introductory plain statement in "When icicles hang by the wall,"[37] but builds up the feeling of winter line by line of varied action and passion. His Winter "maintained by the owl" does not silence its "merry note" and Keats's winter night with its silent owl is the colder for the recollection of this. An even more obvious example might be Thomas's opening line "A grief ago."[38] One has only to substitute "Ten hours" or "Five days" for "A grief" to realize how meaningless the units of time are in themselves. Sometimes the standard is set, not with respect to some differing referent, but from within its own terms, as in Wilde's "And that each day is like a year, / A year whose days are long."[39]

The logistical model of population growth which is embodied in the equation $x_n = g(1 - x_{n-1})x_{n-1}$ is fecund with the most interesting behavior. It has the trivial steady state corresponding to extinction $x_n = 0$. By a steady state we mean one that does not change from year to year, that is, a solution of the equation for which $x_n = x_{n-1} = x$. Thus x must satisfy $x = g(1 - x)x$ and of this equation $x = 0$ is always a solution. Moreover it is not hard to see that it is asymptotically stable for $g < 1$ and unstable for $g > 1$. To see this we have only to remark that the growth parameter $a = g(1 - x)$, where $(1 - x)$ is certainly less than 1 so that $a < 1$ if $g < 1$. $a < 1$, was the condition for stability. Likewise if $g > 1$ then $a = g(1 - x) > 1$ if x is small enough and this is the condition for instability. The trivial steady state therefore has the same properties as in the linear model, but, in contrast to the linear model, the logistical has nontrivial steady states when $g > 1$. To see this we notice that if x is not zero we can divide both sides of the equation for the steady state, $x = g(1 - x)x$, by x and get $1 = g(1 -$

x) or $x = 1 - (1/g)$. For $g > 1$ this nontrivial steady state increases with the increase of the characteristic growth parameter g. For example, if $g = 1.5$, $x = 1/3$; if $g = 2$, $x = 1/2$; if $g = 3$, $x = 2/3$; if $g = 4$, $x = 3/4$.[40]

It is less elementary to determine the stability of the nontrivial steady state and I will not attempt to describe the theory, delightful though it be, but be content with a bald statement of results. It turns out that the nontrivial steady state is asymptotically stable only for $1 < g < 3$ and is unstable for $3 < g < 4$. This means that stable steady populations are always less than two thirds of the saturation population, $x < 2/3$. Such steady populations, obtained for $1 < g < 3$, are asymptotically stable and if the population is artificially perturbed it will, after a few generations, go back to where it was before. Moreover it is not parametrically sensitive to small changes in the characteristic growth parameter g. It will be recalled that the nontrivial steady state of the linear model was sensitive to the growth parameter a, the slightest increase or decrease of which led either to explosion or to extinction. By contrast, the nontrivial steady state of the logistic model will change slightly if g is changed slightly, but will not behave catastrophically. Thus if $g = 2$ so that the nontrivial steady state were $x = 0.5$, a small change of g, say to 2.1, would change the steady state population to $x = 0.524$, but the situation would still be quite stable. Only when $g > 3$ would something radically different happen if the nontrivial steady state population were perturbed. For example, if $g = 3.1$, the nontrivial steady state would be $x = 0.677$, but the slightest perturbation, say to 0.68, would lead, not to unlimited growth or extinction, but to a two-year cycle in the population in which in odd years it would be about 0.765 and in even years about 0.558. Moreover, this two-year cycle would be stable in the sense that any artificial perturbation would die out and the population would soon get back to alternating between 0.765 and 0.558. What's more, it would not be unduly sensitive to small changes in g. If g were 3.15 the numbers 0.765 and 0.558 would change slightly to 0.783 and 0.533, but it would still be a stable two-year cycle.

As if this development of a two-year cycle were not dramatic enough, the two-year cycle loses its stability when $g > 3.45$ and the only stable behavior is a four-year cycle. As the characteristic growth parameter g is increased still further this breaks down and a stable eight-year cycle obtains. This goes on with 16, 32, 64 . . . -year cycles being generated more and more rapidly, until a characteristic growth rate is reached at which no cyclic behavior of whatever period is stable. From this point onward,

Chaos Umpire sits,
And by decision more embroiles the fray
By which he Reigns:[41]

Indeed this behavior is known to mathematicians as "chaos." It is deterministic, for it contains no random element and, as in Milton, is to be distinguished from the "high Arbiter *Chance*." If chance ruled, it would not be possible to calculate the population x_n from x_{n-1}, but it is possible—and by a very simple formula too. So, though perfectly predictable, the population fluctuates quite irregularly and never settles down to a steady or cyclic behavior. The population histories of two populations that start extremely close together will, after a sufficient length of time, bear absolutely no relationship to one another. There are some rather subtle traces of order in this chaos, but they are not easy to describe and I will not attempt to take the description of the model further.[42] There is also an exquisite universality about the model which allows us to replace the growth law $g(1 - x)x$ with any function having the same general shape and be assured of getting qualitatively the same behavior.[43]

I have described the population model in some detail as being perhaps less familiar than the numerous poems that the reader will be able to call to mind, and hope that the minimal algebraic detail that is necessary will not have been too offputting. In drawing attention to certain features, such as the importance of dimensionless variables, that seem to me to have echoes in poetic principle and craft, I want to stress that they will often not bear the full weight of analogy, still less of isomorphism. Isomorphism and homeomorphism are precise concepts applicable to the tightly defined systems of mathematics. It would be a mistake to apply them to Keats's and Bridges's nightingales or Wordsworth's and Shelley's skylarks. Of course, echoes abound in the very words of all subjects—"text"[44] and "fiber bundle,"[45] for instance, are both redolent of the weaver's shop—and such echoes are a constant source of delight to us all, but the *vincula* that I am seeking go beneath the verbal surface to the mode of thought that is being employed. There are other links which I am sure someone less painfully limited to his own immediate experience in literature and mathematics than I could tease out.

There is a certain dynamic tension when two ideas at something of an angle to one another are let loose in a poem such as Hopkins's "That Nature is a Heraclitian Fire and of the comfort of the Resurrection." They do not have the complete orthogonality of oxymoron, but are sufficiently disparate to allow still stranger elements to be ordered into a certain harmony and coinhere in a movement toward the climax:

In a flash, at a trumpet crash,
I am all at once what Christ is, since he was what I am, and
This Jack, joke, poor potsherd, patch, matchwood, immortal
diamond,
Is immortal diamond.

Similar tensions can be found in a model of a honeycomb structure that was used to define an equivalent uniform medium of the same thermal properties. The hexagons that represent the cells of the honeycomb structure require three quantities to define them (their opposite sides being equal), but they fill a space of only two dimensions. This implies that all formulae must have both the two dimensional character of the plane and the threefold symmetry of the hexagon and leads to a final result that embodies this tension.[46]

Then there is the common struggle to make the words and concepts of a poem or model crystalline in their character and suitability. The mathematician will instantly empathize with Eliot. He too is

Trying to learn to use words, and every attempt
Is a wholly new start, and a different kind of failure
. . . And so each venture
Is a new beginning, a raid on the inarticulate
With shabby equipment. . . .[47]

Like Eliot, he finds his means of expression

strain,
Crack and sometimes break, under the burden,
Under the tension, slip, slide, perish,
Decay with imprecision, will not stay in place,
Will not stay still.[48]

Finally (for was not the last principle of Dryden's *Parallel* the knowledge of when to "give over"?), the iterative character of mathematical modeling corresponds very much to the thought expressed by Eliot in "Little Gidding":

What we call the beginning is often the end
And to make an end is to make a beginning
The end is where we start from.[49]

Or again: "Every phrase and every sentence is an end and a beginning, / Every poem an epitaph."[50] Similarly every stage of a mathematical model can often be the starting point for further developments. The model itself may be part of a cyclic process in which the model leads to experience whose analysis leads to further development of the model tending to greater fidelity of representation or greater clarity of concept.[51]

Nothing could better express the aspirations of the natural philosopher in general or the peculiar efforts of the mathematical modeler in particular than Eliot's lines in the concluding section of *Four Quartets:*

> We shall not cease from exploration
> And the end of all our exploring
> Will be to arrive where we started
> And know the place for the first time.[52]

UNIVERSITY OF MINNESOTA

NOTES

1 Cicero *Pro A. Licinio Archia Poeta Oratio ad Iudices* 1.2.

2 William Jackson, *Thirty Letters on various subjects,* 2nd ed. (1784; rpt. New York, 1970). The whole of the passage, which is quoted by Ralph Cohen at the head of his chapter on the scope of critical analogy in *The Art of Discrimination* (Berkeley, 1964), p. 188, reads as follows: "Is there not something very fanciful in the analogy that some people have discovered between the arts? I do not deny the *commune quoddam vinculum,* but would keep the principle within its proper bounds. Poetry and painting, I believe, are only allied to music and to each other; but music, besides having the above-named ladies for sisters, has also astronomy and geometry for brothers, and grammar—for cousin, at least."

3 Horace *De Arte Poetica* 361; Jean H. Hagstrum, *The Sister Arts* (Chicago, 1958), p. 3.

4 V. Sackville-West, *The Land: Summer,* in *Collected Poems* (New York, 1934), p. 86. The continuation of this passage brings out the analogy very well:

> Control is theirs. They have ignored the subtle
> Release of spirit from the jail of shape.
> They have been concerned with prison, not escape;
> Pinioned the fact, and let the rest go free,
> And out of need made inadvertent art.
> All things designed to play a faithful part
> Build up their plain particular poetry.
> Tools have their own integrity;
> The sneath of scythe curves rightly to the hand,
> The hammer knows its balance, knife its edge,
> All tools inevitably planned,
> Stout friends with pledge

Of service; with their crochets too
That masters understand,
And proper character, and separate heart,
But always to their chosen temper true.
—So language, smithied at the common fire,
Grew to its use; as sneath and shank and haft
Of well-grained wood, nice instruments of craft,
Curve to the simple mould the hands require,
Born of the needs of man.
The poet like the artisan
Works lonely with his tools; picks up each one,
Blunt mallet knowing, and the quick thin blade,
And plane that travels when the hewing's done;
Rejects and chooses; scores a fresh faint line;
Sharpens, intent upon his chiselling;
Bends lower to examine his design,
If it be truly made,
And brings perfection to so slight a thing.

(pp. 86–87)

5 William Wordsworth, Preface to *Lyrical Ballads* (1802), ed. R. L. Brett and A. R. Jones (London, 1963), pp. 252, 253. Wordsworth further notes that "the Poet, prompted by this feeling of pleasure which accompanies him through the whole course of his studies, converses with general nature with affections akin to those, which, through labour and length of time, the Man of Science has raised up in himself, by conversing with those particular parts of nature which are the objects of his studies. The knowledge both of the Poet and the Man of Science is pleasure; but the knowledge of the one cleaves to us as a necessary part of our existence, our natural and unalienable inheritance; the other is a personal and individual acquisition, slow to come to us, and by no habitual and direct sympathy connecting us with our fellow-beings. The Man of Science seeks truth as a remote and unknown benefactor; he cherishes and loves it in his solitude: the Poet, singing a song in which all human beings join with him, rejoices in the presence of truth as our visible friend and hourly companion" (p. 253). (I owe the reference to this passage [and, indeed, many other good things] to my colleague Tom Clayton.)

6 Plutarch *De gloria Atheniensium* 3. (346 F).

7 [Cicero] *Ad Herennium* 4.39.

8 C. O. Brink, *The "Ars Poetica,"* Vol. II of *Horace on Poetry* (Cambridge, 1971), p. 370.

9 Gotthold Ephraim Lessing, Preface, *Laokoön:* (1766), tr. and ed. Dorothy Reich (Oxford, 1965); Irving Babbitt, *The New Laokoön: An Essay on the Confusion of the Arts* (Boston, 1910). The bibliography is immense, but the following may be mentioned in addition to the books of Cohen and Hagstrum: Mario Praz, *Mnemosyne: The Parallel between Literature and the Visual Arts* (Princeton, 1970); R. W. Lee, " 'Ut Pictura Poesis': The Humanistic Theory of Painting," *Art Bulletin,* 22 (1940), 197–269; John Graham, "Ut Pictura Poesis," *Dictionary of the History of Ideas* (New York, 1973); and Anthony Blunt, *Artistic Theory in Italy 1450–1600* (Oxford, 1940).

10 See W. Folkierski, "Ut pictura poesis ou l'étrange fortune du *De arte graphica* de Du Fresnoy en Angleterre," *Revue de Littérature comparée,* 27 (1953), 385–402.

11 Hagstrum, p. 175.

12 William Mason, "The Life of M. du Fresnoy," in *The Art of Painting of Charles Alphonse Du Fresnoy* (Dublin, 1783), pp. xx–xxi.

13 John Dryden, *The Art of Painting by C. A. Du Fresnoy* (London, 1695). It was reissued in 1716 by Richard Graham, prefaced by Pope's Epistle to Jervas. Other translators were Defoe (1720), Wright (1728), and Wells (1765). Johnson drew on it to illustrate artistic terms in his dictionary.

14 He confesses at the end it was "begun and ended in twelve mornings" and quotes Rochester's comment on a tragedy commended to him for having been written in only three weeks: "How the Devil could he be so long about it? for that Poem was infamously bad." See John Dryden, *A Parallel of Poetry and Painting*, in *The Literary works of Sir Joshua Reynolds*, Vol. III (London, 1819), p. 281.

15 William Mason, whose translation had the benefit of notes by Sir Joshua Reynolds, also printed Pope's *Letter to Mr Jervas* (1716) "in order that a Poem which does so much honour to the original author may still accompany his work, although the translator [i.e., Mason] is but too conscious how much so masterly a piece of versification on the subject of Painting, will, by being brought thus near it, prejudice his own lines." Mason also wrote an *Epistle to Sir Joshua Reynolds* in the hope of "obviat[ing] . . . every suspicion of arrogance in attempting this work after Mr. Dryden" which has the lines,

> His pen in haste the hireling task to close
> Transform'd the studied strain to careless prose,
> Which, fondly lending faith to French pretence,
> Mistook its meaning, or obscur'd its sense.
>
> Yet still he pleas'd, for *Dryden* still must please,
> Whether with artless elegance and ease
> He glides in prose, or from its tinkling chime,
> By varied pauses, purifies his rhyme,
> And mounts on Maro's plumes, and soars his heights sublime.

All these are conveniently to be found in *The Literary Works of Sir Joshua Reynolds*, Vol. III, pp. 284, 11, 3–4.

16 See Hagstrum, p. 60, n. 14, where the various editions are listed. The 1711 edition of Bentley seems to have settled the question for subsequent editors. On ancient variants and the "Acronian" scholia see Brink, pp. 35–43.

17 For a variety of translations of Horace's "ut pictura poesis," see Cohen, p. 190 n.

18 Horace, 363. In Brink the line is given as "haec amat obscurum, uolet haec sub luce videri."

19 Southey uses the pictorial analogy in his criticism of *Lyrical Ballads* when he likens "The Idiot Boy" to "a Flemish picture in the worthlessness of its design and the excellence of its execution." See Robert Southey, rev. of *Lyrical Ballads, The Critical Review*, 24 (1798); as quoted in app. C of *Lyrical Ballads: Wordsworth and Coleridge*, ed. R. L. Brett and A. R. Jones (London, 1963), p. 319.

20 It should be emphasized that this in no way does justice to the scope of critical comment that comes under the rubric of "Ut pictura poesis." For this one needs to turn to Cohen's *The Art of Discrimination*, esp. ch. 4, pp. 188–215.

21 Mischa Penn and Rutherford Aris, "The Mere Notion of a Model," *Mathematical Modelling*, 1 (1980), 1–12.

22 Robert Bridges, *Shorter Poems* (Oxford, 1953), Bk. IV. It is interesting that the now obsolete (except perhaps in Scotland) usage of "maker" for "poet" translates rather than transliterates the Greek.

23 Michael Riffaterre, *Semiotics of Poetry* (Bloomington, Ind., 1978): "Poetry expresses concepts and things by indirection. To put it simply a poem says one thing and means another" (p. 30).

24 I cannot refrain from remarking that, as a tolerably educated person, with no colorable claim to being really well read by the standards that should prevail in a university but who has read a good deal of literary theory and criticism during the last year or two, I have been struck, not by its vagueness, but by the passion with which much of the modern sort espouses a particular viewpoint. No doubt our "class" background and ideology colors our reading and writing, but to make it the sole, or even the dominant, determinant seems unnecessarily perverse. Feminist critics have given, and are giving, us a valuable corrective to past blindness, but feminism, like masculism, can at best be a half-way house to humanism. Freud and Lévi-Strauss are pivotal figures in modern thought, but must everything be suppression or opposition? In some cases I would have thought they were having me on were it not that they showed no glimmer of that sense of humor which is the grace of the scholar, much as a holy hilarity is of the saint. The only parallel I know of in mathematics is the competition between certain numerical analysts, some of whom espouse finite differences whilst others are wedded to finite elements.

25 See Graham Goulden Hough, *An Essay on Criticism* (New York, 1966): "The point is not, or ought not to be, that literature 'imitates' objects in the real world: so does scientific and historical writing. The point is that literature creates fictitious objects" (p. 42).

26 Maynard Smith, in *Models in Ecology* (Cambridge, 1974), has suggested that the word "simulation" be used in this case to leave "model" for the broader case of clarifying concepts.

27 Robert Bridges, *The Testament of Beauty* (Oxford, 1929), I, 367. The spelling is Bridges's.

28 Godfrey Harold Hardy, *A Mathematician's Apology* (Cambridge, 1940), p. 80.

29 Leonard Welsted, *Epistles, Odes, &c.* (London, 1728), quoted by Cohen, p. 76.

30 I think it is this sense of imaginative oneness that Sir Frederick Pollock had in mind in his biographical memoir of his friend, the mathematician W. K. Clifford, where he wrote, "It is an open secret to those who know it, but a mystery and a stumbling-block to the many, that Science and Poetry are own sisters" (quoted in John Pollock, *Time's Chariot* [London, 1950]). That the less imaginative aspect of science, its accuracy of observation and definiteness of knowledge, can inform poetic expression and criticism has been well illustrated by Roy Broadbent Fuller in his lecture on the relationship of science and poetry, "The Osmotic Sap," in *Professors and Gods* (London, 1973), pp. 28–46. He deals with larger questions than have been essayed here, in particular with the role of science as the informing influence of its time. He contrasts the influence of the scientific temper in the thirties on the direction of literature with the non- or antiscientific attitude in the fifties and sixties. He refers also to I. A. Richards's *Science and Poetry*, 2nd rev. ed. (London, 1935).

31 p_n is said to be a function of p_{n-1}, and this is written $p_n = f(p_{n-1})$.

32 For example, MS Bodley 764, f.12r. Cf. British Library, Harley 4751.

33 The constant, *a*, must, of course, be a positive number since we could not have a negative population.

34 Jon. 2:5.

35 A. E. Housman, Preface, *M. Manilii Astronomicon Liber Quintus*, ed. A. E. Housman (London, 1930), pp. xxxv–xxxvi. De la Mare's word was "rusting." Another example is Housman's criticism of Swinburne's immoderate praise of Shelley's "exquisite inequality" in his Inaugural of 1911 (published as *The Confines of Criticism*, ed. John Carter [Cambridge, 1969]).

36 F. W. Bateson, "The Poetry of Emphasis," in *A. E. Housman: A Collection of Critical Essays*, ed. Christopher Ricks (Englewood Cliffs, N.J., 1968), pp. 130–45, points out that Housman missed the other misprint, namely "these" for "those" in line 3. "Those"

might be surmised as it balances "this my dust" in line 1, but this is a logical, rather than a pictorial, sensibility and tells us (as Bateson would have it) something about Housman's "emphasis."

37 William Shakespeare, *Love's Labour's Lost,* 5.2.920.

38 Dylan Thomas, *Collected Poems* (New York, 1957), p. 54.

39 Oscar Wilde, *The Ballad of Reading Gaol* (1898), in *The Portable Oscar Wilde,* ed. Richard Aldington and Stanley Weintraub (Harmondsworth, 1981), p. 685.

40 We do not consider $g > 4$ since then x could be greater than 1 or even negative.

41 John Milton, *Paradise Lost,* in *John Milton: Complete Poems and Major Prose,* ed. Merritt Y. Hughes (Indianapolis, 1957), Bk. II, l. 907.

42 A recent conference on these matters, held in Los Alamos, N. Mex., May 1984, had the engaging title "Kosmos en Chao."

43 For a fuller, but still elementary, account of this model see my "Re, k and pi; a conversation on some aspects of mathematical modelling," *Applied Mathematical Modelling,* 1 (1977), 386.

44 "Verba eadem qua compositione vel in textu iungantur vel claudantur." See, e.g., Quintillian *Inst.* 9.4.13.

45 A term in the branch of mathematics known as topology.

46 Rutherford Aris, "De exemplo simulacrorum continuorum discretalumque," *Archive for Rational Mechanics and Analysis,* 70 (1979), 203. A translation is to be found in Appendix C of my *Chemical Engineering in a University Context* (Madison, 1982).

47 T. S. Eliot, "East Coker," in *Four Quartets* (New York, 1943), 11. 174–75, 178–80. The mathematician will empathize too with the way in which small errors can creep in, as when a little later on in this section of "East Coker," the false concord of "Here and there does not matter" (1. 203) was overlooked in proof and stood until 1974. See Helen Louise Gardner, *The Composition of Four Quartets* (London, 1978), p. 113.

48 T. S. Eliot, "Burnt Norton," in *Four Quartets,* 11. 150–54.

49 T. S. Eliot, "Little Gidding," in *Four Quartets,* 1. 214. The relationship of end and beginning is a continually recurring theme in *Four Quartets.* In "East Coker" (11. 1 and 14) there is an almost direct paraphrase of Guillaume de Machaut's cryptic rondeau "Ma fin est mon commencement / et mon commencement ma fin," but whether this is a conscious allusion is, as so often with Eliot, hard to tell.

50 "Little Gidding," 1. 224.

51 See n. 25.

52 "Little Gidding," 1. 239.

FIFTH P. V. DANCKWERTS MEMORIAL LECTURE PRESENTED AT GLAZIERS' HALL, LONDON, U.K.
16 OCTOBER 1990

MANNERS MAKYTH MODELLERS

RUTHERFORD ARIS
Department of Chemical Engineering and Materials Science, University of Minnesota, Minneapolis, MN 55455, U.S.A.

"What we need" said the Professor, carrying his tea-cup to the table next to his accustomed chair, "is more good research on crushing and grinding."

It was 11 o'clock of a Cambridge morning in that little room of the Shell Chemical Engineering Laboratory which is dedicated to the academic staff's morning tea-break—an institution that Danckwerts once commended in print to his transatlantic colleagues (Danckwerts, 1981, p. 217). As I recall it, few had as yet come in for tea and perhaps that was why, in spite of always feeling a little like a New Boy in the presence of a School Prefect, I thought it incumbent on me to make some reply. So I said something to the effect that I supposed it would be a stochastic process and that one ought to be able to get a Fokker–Planck equation for the evolution of the size distribution. When I subsided, the word came down from Olympus. "What I rather had in mind was some good experimental research on crushing and grinding." There was more discussion in which others took part and I do not recall if there was any consensus as to the need for more good research on crushing and grinding, but the exchange was typical of tea-room conversations.

Shortly after J. D. Murray's analysis of fluidization as two interpenetrating continua had been published in JFM, I could not resist saying that, at last, it would be possible to understand fluidization. This was a bit of cheek on my part, as Davidson and Harrison's book was out and their fruitful collaboration was in full flood. (When Peter Rowe visited the department at about this time and Davidson and Harrison brought him into tea, I remember Robin Turner saying to me that a terrorist bomb at that moment could have wiped out half the knowledge of fluidization in the world.)

I recall these interactions with the great man himself because I think it meet, on such an occasion as this, not to overlook the personal side of the man whose memory we honour in this Danckwerts Lecture. He was the Compleat Chemical Engineer, rooted in Oxford chemistry, a product of MIT's Practice School, doyen of the Cambridge School, Professor in the days when there were few such chairs, President of the Institution. The selection of his papers, which we are fortunate enough to have with his comments, shows the scope of his work: diffusion problems, gas absorption, the reaction of carbon dioxide in aqueous solutions, residence time distributions and mixing.

He had the engineer's mistrust of too much mathematics. "I have felt for some years," he said in a review of the historical survey "A Century of Chemical Engineering" (Danckwerts, 1982), "that chemical engineering is weighted-down with more mathematics than it can support." But—and in this feature alone was he like Betjeman's Wykehamist—he was "broad in mind" and acknowledged that the history of the use of mathematics in chemical engineering was of great interest.

But it was an intelligent mistrust, not an ignorant aversion, and he knew when mathematics was needed and how to use it, so that there was challenge, rather than contempt, in his animadversions upon mathematical modelling. It would be unfair to call it an affectation, for affectations are superficial and Danckwerts was not a superficial man. It was of a piece with the remarks (*Insights*, pp. ix and x) that he (and Winston Churchill) made on the amount of time they had to waste on Latin prose composition during their schooldays—he (and Churchill) knew well enough that nothing is better either for English style or for general discipline of the mind, but the small boy in him could not pass that particular house without rattling a stick across its iron fence. It was a minor foible, a mannerism perhaps, the small change of manners. For by manners had he been made, as one expects in a Wykehamist and I need hardly say that I intend no derogation of England's premier school by having adapted its motto in my title.

"Manners" is a word of rich overtones, embracing much more than the decencies of social intercourse. It is the whole way in which a thing is done, how the hand (*manus*) is put to the plough, and hence the mode of one's habitual behaviour and conduct, especially in its moral aspect; it can be genetically individual ["but to my mind,—though I am native here,/And to the manner born,—it is a custom/More honour'd in the breach than the observance" (*Hamlet* I, iv, 14)] or characteristically public ("Within this hour it will be dinner time:/Till that, I'll view the manners of the town,/Peruse the traders, gaze upon the buildings" (*Comedy of Errors* I, ii, 12)]. It has all manner of meanings and shades of meaning and a contemporary critic would not need to invent any spurious connotations in order to deconstruct them. Doubtless there are many layers to the famous motto "Manners makyth man" [cf. Firth (1936)], but this is not the place to peel them back. What I hope is conveyed by my title is that there are certain characteristic styles in the use of mathematical models in chemical engineering and that their several virtues and limitations will bear looking into.

WHAT IS A MATHEMATICAL MODEL?

First it should be said that the term "mathematical modelling" took a tremendous leap in popularity some 15 years ago. There are now journals of mathematical modelling in various forms and an International Society to canonize an activity that has been going on for many years. "Par ma foi!", we are apt to exclaim with M. Jourdain, "il y a plus de quarante ans que je dit de la prose (or *fait des modèles mathematiques*) sans que j'en susse rien." For mathematical models come in all sizes and some of the best of them, for example, the model underlying Danckwerts' theory of residence time distributions, involve only elementary mathematical considerations.

It is an essential quality in a model that it should be capable of having a life of its own. It may not, in practice, need to be sundered from its physical matrix. It may be a poor thing, an ill-favoured thing when it is by itself. But it must be capable of having this independence. Thus Liljenroth, in his seminal paper on multiplicity of steady states, can hardly be said to have a mathematical model, unless a graphical representation of the case is a model. He works out the slope of the heat removal line from the ratio of numerical values of a heat of reaction and a heat capacity. Certainly he is dealing with a typical case, and his conclusions are meant to have application beyond this particularity, but the mechanism for doing this is not there. To say this is not to detract from Liljenroth's paper, which is a landmark of the chemical engineering literature; it is just to notice a matter of style and the point at which a mathematical model is born. For in the next papers on the question of multiple steady states, those of Wagner (1945), Denbigh (1944, 1947), Denbigh *et al.* (1948) and van Heerden (1953), we do find more general structures. How powerful the life that is instinct in a true mathematical model can be seen from the Fourier's theory of heat conduction, where the mathematical equations are fecund of all manner of purely mathematical developments.

At the other end of the scale a model can cease to be a model by becoming too large and too detailed a simulation of a situation whose natural line of development is to the particular rather than to the general. It ceases to have a life of its own by becoming dependent for its vitality on its physical realization. Maynard Smith (1974) was, I believe, the first to draw the distinction in ecological models between those that aimed at predicting the population level with greater and greater accuracy (simulations) and those that seek to disentangle the factors that affect population growth in a more general way (models). The distinction is not a hard and fast one, but it is useful to discern these alternatives.

Though the model may have a life of its own, it does not exist in isolation, nor provide all the answers of itself. It is better thought of as part of a process of understanding in which a preliminary concept leads to a first model on the basis of which experiments can be designed. The experience that this generates must be evaluated and interpreted and leads to revised ideas on the basis of which an improved model can be constructed. This is a cyclic process which may lead to greater conceptual clarity of the situation in general or deeper understanding of a particular case.

The level of mathematical sophistication is, of course, an index of style or manner and one could make a spectrum of chemical engineering modellers from Amundson to Zygourakis, slotting each into their mathematical wavelength. I will not do so, not merely because it would be the quickest way to lose friends and not influence the subject, but because it is a superficial classification. For one thing, it is a function of time. Amundson was being sophisticated as a chemical engineer when he used matrices in a 1946 paper on distillation; today he would not need to preface his paper with an elementary introduction to matrix algebra. It is not sophistication but simplification that is the hallmark of mathematics, "thatt supersensuous sublimation of thought, the euristic vision of mathematical trance", as Bridges calls it. It is the penetration of mathematical thought that is to be coveted; its ability to get through to the essential form. It is the realization that stoicheiometry has the same structure as linear algebra, or that a chemical reactor is a reification of a dynamical system, or that the extremes of complete mixing and plug flow can be united, and indeed many more cases subsumed under the same formulation, by the concept of the residence time distribution.

In the context of mathematical modelling Bridges' insight is certainly true and germane. He is concerned in the section of Book I of *The Testament of Beauty* with the tendency to divorce human thought from natural phenomena, to set man outwith the house of nature into which he peers "hooding his eyes" to see what is within rather than his own reflection. He goes on in lines 358–370:

> See how they hav made o' the window an impermeable wall
> partitioning man off from the rest of nature
> with stronger impertinence than Science can allow.
> Man's mind, Nature's encrusted gem, her own mirror
> cannot bë isolated from her other works
> by self-abstraction of its unique fecundity
> in the new realm of his transcendent life;—

operation (it is a matter of multiplying by a probability and integrating over all possibilities) and since C appears linearly in $\langle c \rangle$, the expected reaction rate must be proportional to $[C]$, the average concentration in the fluid phase, because the well-mixed particles have been everywhere and been exposed to every concentration in the bed with equal probability. After all that is what is meant by being "well-mixed".

To put this in another manner, let $P(t;\ z;\ C',\ t')\,dz\,dC'$ be the probability that a particle of age t is in the slice $(z, z + dz)$ and that when its age was t' it was exposed to a concentration in the range $(C', C' + dC')$. Again let $p(t;\ z;\ z',\ z',\ t')\,dz\,dz'$ be the probability that just such a particle was in the interval $(z', z' + dz')$ at time t'. Clearly, if the particle is moving through a steady concentration field, $C(z)$, then

$$P(t;\ z;\ C',\ t')\,dz\,dC' = p(t;\ z;\ z',\ t')\,dz\,dz' \quad (10)$$

if $C' = C(z')$. By "being well-mixed" we mean that p is independent of both z and z', since a particle can be anywhere at any time with equal probability. Thus $p = 1/L^2$. The average concentration in particles of age t is therefore independent of position (again that is the meaning of "well-mixed") and is

$$\begin{aligned}\int P(t;\ z;\ C(t'),\ t']\,dC \int_0^t C(t')F(t - t')\,dt' &= \int_0^L p(t;\ z;\ z',\ t')C(z')\,dz' \int_0^t F(t - t')\,dt' \\ &= [C]L^{-2} \int_0^t F(t')\,dt'.\end{aligned}$$

The last step uses the constancy of p and a change of variable from t' to $t - t'$. To get R, which we now see must be independent of z, we have to average over all ages. Let ε and $(1 - \varepsilon)$ be the fractions of the reactor volume occupied by the fluid and solid phases, respectively; then εAL is the volume of catalyst in a bed of area A and length L, and, if q_s is the rate at which it is added and removed, $\theta_s = \varepsilon AL/q_s$ is the mean residence time of the catalyst and its age distribution is $\exp - (t/\theta_s)/\theta_s$. Thus $R = \alpha[C]$, where

$$\begin{aligned}\varepsilon\alpha &= (1 - \varepsilon)k \int \exp - (t/\theta_s)\,dt/\theta_s \int F(t')\,dt' \\ &= (1 - \varepsilon)k\theta_s f(1/\theta_s) \quad (11)\end{aligned}$$

and $f(s)$ is the Laplace transform of $F(t)$. It is an unexpected benefit of the exponential age distribution that the inversion of the Laplace transform is not required.

We can now make a balance of A* in the segment of the reactor between z and $z + dz$, equating the net flow into the segment in the fluid phase to the rate of reaction and the withdrawal rate of the solid. Thus

$$\varepsilon A\{D^* C'' - vC'\}\,dz = \varepsilon\alpha[C]A\,dz + q_s(dz/L)(\alpha/k)[C] \quad (12)$$

or

$$P^{-1}C'' - C' = \beta[C] \quad (13)$$

where

$$\beta = \alpha(1 + k\theta_s)/k\theta_s \quad (14)$$

and the prime denotes the derivative with respect to $x = z/L$. The last term arises from the fact that no A* is being fed to the reactor with the solid whereas it is being withdrawn at the average particle concentration. This equation is subject to

$$- P^{-1}C'(0) + C(0) = C_f \text{ and } C'(L) = 0. \quad (15)$$

Solving the differential equation is easy, since $[C]$ is a constant which we can determine afterwards. The solution is

$$C(x) = C_f - \beta[C]\{1 + Px - \exp - P(1 - x)\}/P. \quad (16)$$

Averaging gives

$$[C] = C_f/\{1 + \beta\chi(P)\} \quad (17)$$

where

$$\chi(P) = \{P^2 + 2P - 2 + 2\exp - P\}/2P^2. \quad (18)$$

We can now substitute $[C]$ in $C(x)$ and set $x = 1$ to get the exit concentration, namely

$$C(1) = C_f\{1 - \beta(1 - \chi)\}/\{1 + \beta\chi\}. \quad (19)$$

Now χ varies from 1 to 0.5 as P varies from 0 to infinity, so $1 - \chi$ is a positive number, zero only when $P = 0$. But β can be made as large as we please quite independently of χ which depends only on P, whereas β depends on the other parameters. As soon as $\beta > (1 - \chi)^{-1}$, $C(1)$ is negative! The only reprieve would come if β were always less than 2. But

$$\beta = \alpha(1 + k\theta_s)/k\theta_s = \{(1 - \varepsilon)/\varepsilon\}[(1 + k\theta_s)f(1/\theta_s)] \quad (20)$$

and even though the expression in the square bracket is bounded, the ratio $(1 - \varepsilon)/\varepsilon$ can be made as large as we please by flooding the system with catalyst. This choice can be made independently of the other properties. Though, if we insist on thinking of this as a fluid bed, ε cannot be made arbitrarily small.

A SIMPLER SYSTEM

However a simpler system, much less suspect in its assumptions, suggests itself. Let the reactor consist of a tube of length L and volume εV through which there is flow with dispersion, but no reaction. The reaction takes place in a stirred tank of volume $V(1 - \varepsilon)$ with which the tube communicates through its porous wall. With the experience of the previous problem the equations can be written down with an obvious notation:

$$\varepsilon\{D^*(d^2C/dz^2) - v(dC/dz)\} = (1 - \varepsilon)kc + (q/V)c. \quad (21)$$

The last term again arises from the fact that no A* flows into the stirred tank reactor whereas it flows out at a concentration c in the effluent stream of volumetric rate q. The equation for the stirred tank is

$$(Sk_c/L)\int_0^L (C - c)\,dz = (1 - \varepsilon)Vkc + qc. \quad (22)$$

Introducing the dimensionless quantities

$$U = C/C_f,\ u = c/C_f,\ x = z/L$$

$$m = Sk_c/q,\ n = (1 - \varepsilon)Vk/q,\ P = vL/D^*,\ b = qL/V\varepsilon v$$

gives the equations

$$(1/P)U'' - U' = b(1 + n)u \quad (23)$$

$$m[U] = (1 + m + n)u \quad (24)$$

with

$$-U'(0)/P + U(0) = 1 \text{ and } U'(1) = 0. \quad (25)$$

Let us look at the solution for P tending to infinity since the difficulty shows up most easily in this case. Then

$$U(x) = 1 - \{bm(1 + n)/(1 + m + n)\}[U]x \quad (26)$$

$$[U] = 1/\{1 + [bm(1 + n)/2(1 + m + n)]\} \quad (27)$$

$$U(1) = \{1 - [bm(1 + n)/2(1 + m + n)]\}/\{1 + [bm(1 + n)/2(1 + m + n)]\}. \quad (28)$$

Once again this can be negative.

But there is another way of tackling this problem. Instead of doing a balance of A* over a section of both the tube and the stirred reactor between z and $z + dz$, in which the transfer term does not appear, let us do it only over the tube. Then

$$\varepsilon\{D^*(d^2C/dz^2) - v(dC/dz)\} = (Sk_c/V)(C - c) \quad (29)$$

or

$$(1/P)U'' - U' = bm(U - u) \quad (30)$$

with the same boundary conditions and the same relation between u and $[U]$. If we again look at the extreme case of infinite P:

$$U(x) = u + (1 - u)\exp - bmx,$$

$$u = m[U]/(1 + m + n) \quad (31)$$

whence

$$[U] = (1 + m + n)E/\{1 + n + mE\},$$

$$E = [1 - \exp - bm]/bm \quad (32)$$

and

$$U(1) = \{(1 + n)e + mE\}/\{1 + n + mE\},$$

$$e = \exp - bm. \quad (33)$$

This is certainly positive and presents none of the problems of the other solution.

THE PUZZLE REVISITED

This simplified model raises a number of questions. Why do the two approaches not agree? Are we to conclude that the invalidity of the result in the first case shows that the method is wrong? Not without other reasons, for it could be wrong at some other point. Are we wrong in making a balance over anything less than the complete stirred tank? I see no *a priori* reason that would forbid this. Have we done the balance correctly? I think so, but would be glad of correction if I have erred.

But wait! Return to the possibility that we were in error by taking a balance over anything less that the whole stirred tank. If it is permissible to take a balance over the $(z, z + dz)$ slice of both reactors, then it is permissible to take balances on the tubular and on the stirred reactor separately. Doing so on the stirred side gives

$$Sk_c(C - c) = (1 - \varepsilon)Vkc + qc. \quad (34)$$

But, since c is independent of z, so also is C and therefore $C' = 0$. Thus the two concentrations must both be everywhere zero and this is inconsistent with the feed condition $C = C_f$. It follows that eq. (23) is invalid and it is eqs (24) and (30) that must be solved. Doing this we find that

$$U(1) = (1 + n)/\{m\mathbf{G} + (1 + n)\mathbf{H}\} \quad (35)$$

where

$$\mathbf{G} = (M/NQ)(e^N - 1) + (N/MQ)(1 - e^{-M})$$

$$\mathbf{H} = (M/Q)(1 + N)e^N + (N/Q)(1 - M)e^{-M}$$

and

$$Q^2 = P^2 + 4Pbm,\ M = (P + Q)/2,\ N = (Q - P)/2. \quad (36)$$

Both these functions are positive and, hence, U is always positive.

Can we apply the second method to the fluidized-bed problem? The expected value of the surface concentration, rather than that of the mean concentration, could be calculated and would be proportional to the average concentration $[C]$. An equation of the form

$$P^{-1}C'' - C' - \psi C = -\psi\gamma[C] \quad (37)$$

would take the place of (13) with

$$\psi = 3(1 - \varepsilon)Lk_c/\varepsilon va,\ \gamma = \mu/[\mu + (\kappa^2/3)\eta(\kappa\varphi)],$$

$$\eta(\sigma) = 3(\sigma \coth \sigma - 1)/\sigma^2. \quad (38)$$

If the two eigenvalues of (37) are denoted by M and $-N$, so that

$$Q^2 = P^2 + 4P\psi,\quad M = (Q + P)/2,\quad N = (Q - P)/2 \quad (39)$$

and we define the functions

$$\Gamma = (M/NQ)(e^N - 1) + (N/MQ)(1 - e^{-M}) = \mathbf{G},$$

$$\Delta = (M^2e^N - N^2e^{-M})/PQ. \quad (40)$$

Then

$$C(1)/C_f = \{\kappa^2\eta(\kappa\varphi)/3 + \mu\Gamma\}/\{\kappa^2\eta(\kappa\varphi)\Delta/3 + \mu\Gamma\} \quad (41)$$

is the exit concentration. This is positive and shows the expected separation of the various parameters. Nor are we surprised to see the Thiele effectiveness factor appearing to account for the internal diffusion limitation.

This seems to resolve the ancient error satisfactorily; a correction of the heat transfer problem will be given elsewhere. The principle at work seems to be that the deliberate mixing, whether of particles or reacting fluid, establishes a forced connection between different parts of the boundary so that on the reactive side they are non-local. In the fluid phase or on the tube side it is legitimate to apply pointwise balances, but not in the solid or stirred-tank side.

SOME MANNERS OF THE CRAFT

A mathematical model should be fully seized of its purpose whether this be synthesis or analysis, design or understanding. Its purpose will shape it to a large extent and certainly size it. If the objective is to develop a structured model of *Saccharomyces cerevisiae* you have to be prepared to have 18 balance equations and 26 rate expressions (Shuler and Steinmeyer, 1989). Nothing much smaller will do the job. If an explanation of chaotic behavior is sought, at least three state variables will be needed. If the purpose is to understand phase separation on a catalytic surface, mean field theory will not do and one may have to have recourse to finite models and Monte-Carlo methods. If the purpose is to explore the range of applicability of a class of model then the shape is given and it becomes a question of seeing what physical situations it illuminates. The danger is the temptation to force a physical situation into an inappropriate mathematical mould, but, if this is avoided, the exploration can be quite fruitful and the demands of the physical situation lead to an extension of the mathematical technique. Iordache's cultivation of polystochastic models is an example (Iordache, 1987).

Of dimensionless variables I shall have something to say in a moment, but chemical engineers are particularly fortunate to be conversant with such a wide range of dimensionless parameters and to have absorbed, perhaps to a greater degree than other engineers, the beautiful notion that expresses the magnitudes of the quantities of a problem in the problem's own terms. I well remember the first time I grasped the significance of making the variables dimensionless when C. H. Bosanquet showed me that the equations of a conduction problem became universal by taking the time to be Dt/a^2. For the mathematician, all variables are dimensionless and C. H. Bosanquet had a brother L. S. whom he described as a mathematician so pure that if you give a number a meaning he won't touch it! Fortunately dimensionless numbers do have meaning. The adroit choice of dimensionless variables and parameters and the point at which they are introduced are characteristic manners of the modeller [see Becker (1976) and Aris (1978, pp. 60–68)]. Once they have been introduced the ability to check equations by their dimensional consistency is almost entirely lost, though a ghost of it lingers in the feeling for the consistency of an expression that the modeller develops. They are essential to the examination of limiting cases, another characteristic manner of the mathematical modeller. A history of dimensionless parameters would be rewarding; the only essay toward this that I know of is by Layton (1988).

Manners are an expression of principles and there are principles that govern the reduction to dimensionless form. Truly constant quantities must be used as characteristic quantities. Thus although $[C]$ is a constant and independent of z, it would not do to define a dimensionless variable as $C/[C]$. It can be done, but is likely to cause confusion and C/C_f is greatly to be preferred. A second principle is that, except when there is a single kind of dependent variable, the reduction must be complete. I well remember the consternation I caused in an industrial short course by having the temperature, as well as the concentration, in units of moles per unit volume! If there is only one kind of dependent variable, as C and c above, then it can be left "as having a certain picturesquencess" [I quote from the preface to E. C. Titchmarsh's Theory of Fourier Integrals—"I have done certain problems in heat and mass transfer as I think an analyst" (i.e. a mathematical analyst) "should. I have retained such terms as temperature and heat as having a certain picturesqueness. The reader need not know that these things exist."] The third principle is that the dimensionless parameters should bear the burden of showing the comparative importance of the various terms in the equation. Thus C/C_f is in the range (0, 1), as are c/C_f and z/L, and P and $b(1+n)$ show the importance of the terms they are associated with in eq. (23). It is not possible to reduce the number of parameters further for, though $y = Px$, $a = b(1+n)/P$ would leave only a in eq. (23), P would turn up again in eq. (25) as $U'(P) = 0$. Nor can b and $(1+n)$ be usefully united since $(1+n)$ is needed by itself in eq. (24). $(1+n)$ could be renamed as a single parameter, but in $(1+m+n)$ it serves as a useful reminder that there are three process at work removing A*. If the tube were infinite then $y = Px = vz/D^*$ would be the way to go. Any problem of sufficient magnitude gives scope for the individual expression of manners on the part of the modeller.

A final point in this section, which is not intended to be comprehensive, concerns the colloquy of models. What I mean by this is illustrated above by the way in which the transfer of a heat problem to a reaction context, an isomorphism, suggested another form of reactor in which one of the questionable hypotheses of the model, the perfect mixing of the solid phase, could be made rigorous. This allowed me to see that the error did not lie in the probability argument and, eventually, to find it in the way of making the balance. In this sense the models can fruitfully interact with each other, while each enjoys a life of its own.

CRAFT AND SULLEN ART

In conclusion I cannot forbear to mention the affinity between the very practical craft of mathematical modelling and the art and craft of poetry and indeed with the whole literary and artistic life of the mind. It is not that it needs legitimizing, still less that it should be dressed in pretentious garments and seek a connection or genealogy for reason of intellectual snobbery, but, if we are to retain our wholeness and if we have any enjoyment in the arts, it is worthwhile to preserve the link and let it nurture our work. We do not put down the pen when we take up the T-square. {I have tried elsewhere to express this idea [cf. Aris (1969, 1977, 1982, 1983, 1989)].}

I think the key is in the notion of craftsmanship (Penn and Aris, 1980) and the innate sympathy of craftsmen who, as Vita Sackville-West (1934) says, "share a knowledge", having "compelled the growth of pattern to the patient shuttle" and "out of need made inadvertent art". It is the experience of refining our own models that makes us resonate instantly to her lines:

The poet like the artisan
Works lonely with his tools; picks up each one,
Blunt mallet knowing, and the quick thin blade,
And plane that travels when the hewing's done;
Rejects and chooses; scores a fresh faint line;
Sharpens, intent upon his chiselling;
Bends lower to examine his design,
If it be truly made,
And brings perfection to so slight a thing.

Or to Dylan Thomas'

In my craft or sullen art
Exercised in the still night
When only the moon rages . . .

There would seem to be four aspects of model building which relate themselves to aspects of the poetic craft: aptness, intrinsic standards, internal tension, and iterative nature. By aptness I mean that the model is suited to the purpose for which it is designed. PVD's notion of the residence time distribution is most excellently apt. It is condign to the elucidation of many aspects of what goes on in the general mixed vessel. It does not complicate the situation unnecessarily, nor so oversimplify it as to reduce its content to the point of it's not being useful.

In just such a manner does the poet seek the apt word. The word that does not particularily draw attention to itself or that has been drained of its meaning by hackneyed use; which will not "slip, slide, perish,/decay with imprecision", as Eliot has it. Housman's poetic instincts were aroused when he read Walter de la Mare's "Fare Well" for the first time in a newspaper (for so he tells us in the preface to the edition of Manilius).

Oh, when this my dust surrenders
Hand, foot, lip, to dust again,
May these loved and loving faces
Please other men!

May the rustling harvest hedgerow
Still the Travellers's Joy entwine,
And as happy children gather
Posies once mine.

"I knew in a moment", Housman writes, "that Mr. de la Mare had not written 'rustling' and in another moment I had the true word" which was, of course, 'rusting'. Interestingly enough he missed the other misprint, "these" for "those" in line 3, but that is a question of logical balance rather than of pictorial image.

We can think of aptness on several scales. There is the fine grained scale of the word which has just been illustrated and might correspond to some small, but subtle, detail of mathematics, the choice of variables in an integration for example. Even the choice of a symbol demands a certain sympathy and respect for tradition. On the larger scale the choice of a method for tackling a model or of a framework within which to formulate it—e.g. discrete or continuous, lumped or distributed—can be apt or clumsy. Similarly the poet must choose his form. A Shropshire Lad could not have been written in the metre of An Essay on Criticism.

Poetic imagery exercises something of the same function as the use of dimensionless variables. Each creates a magnitude from within its context, whether of the poem or model, and so invests it with a meaning that it could not have in external units. A Reynolds' number means something; a velocity of 5 dm a second may be large or small depending on the context. So the natural instinct of the modeller is to cast the variables in dimensionless form and so obtain the minimum number of parameters. Similarly the poet calls on the creatures of his context. How banal Keats' "Eve of St. Agnes" would be if it read

St. Agnes Eve—Ah, bitter chill it was!
The temperature was barely minus two.

Instead he chooses an image that literally puts a shiver into the shoulders:

The owl, for all his feathers, was a-cold;.

You can see the owl sitting hump-shouldered on a branch, his feathers roused slightly to trap the air and minimize the heat transfer coefficient. And even then failing as the bitter chill penetrates. You can feel the sharp edges of frost-stiffened blades of grass and see the sheep huddled in their pen in the next lines.

The hare limp'd trembling through the frozen grass,
and silent was the flock in woolly fold.

What I mean by internal tension is less easy to illustrate. It is patent from the title of Hopkins "That Nature is a Heraclitian Fire and of the Comfort of the Resurrection". Here an idea from Greek philosophy is juxtaposed with a doctrine of Christian soteriology. Or in "The Windhover" where Hopkins contrasts the apparent effortless achieve and mastery of the dapple-dawn-drawn falcon with the sheer plod of the plough and the gold-vermillion of embers, spent yet still beautiful in their breakup and recalling the plumage of the kestrel.

The most vivid internal tension I have come across in a mathematical model is the attempt to get a matrix of equivalent thermal diffusivities for an array of hexagonal passages (Aris, 1979). These tessalate the two-dimensional plane yet themselves are defined by three sides. The whole problem must be worked with this "twoness" and "threeness" constantly in mind and the final formula both embodies and resolves the tension.

The iterative nature of model building is sufficiently obvious and has been mentioned earlier. We have seen that it is a way of understanding the natural world by mirroring it in an alternative space. The commutative diagram

$$\begin{array}{ccc} \text{Physical reality} & \rightarrow & \text{Physical behavior} \\ \downarrow & & \downarrow \\ \text{Mathematical reality} & \rightarrow & \text{Mathematical solution} \end{array}$$

expresses the relationship. When it is well constructed a good model is capable of suggesting further questions and greater refinements. Its several parts are beginnings and ends in themselves much as Eliot says of phrases and sentences:

> Every phrase and every sentence is an end and a beginning
> Every poem an epitaph.

But "tools have their own integrity" and we would distort our trade by imagining we could ever produce anything so moving as the finale of Four Quarters, though we can perhaps hum along with the the first few bars of its build-up;

> We shall not cease from exploration
> And the end of all our exploring
> Will be to arrive where we started
> And know the place for the first time.

"Omne tulit punctum qui miscuit utile dulce" wrote Horace, and, if we have taken our subject seriously enough to have mixed the *utile* of application with the *dulce* of understanding, and have taken ourselves not so seriously as to forget that getting there is a good deal more than half the fun, we shall be able to enjoy the place of chemical engineering science among the *artes humaniores* and of mathematical modelling among the various activities that become the chemical engineer.

Acknowledgements—This lecture itself acknowledges what I have learned from Danckwerts' example of the arts of modelling. Clearly, if what I have learned from others were to be fully acknowledged, this section would be as long again as the lecture. I am sure, however, that none will mind being represented by Neal Amundson, to whom, more than anyone, the whole profession is indebted for "the mathematical understanding of chemical engineering systems". I need hardly add that they must not be blamed for my peculiar perversities. I have inflicted the "puzzling problem" on many people during the 26 years since Robin Turner shewed me that there was something amiss. The fact that no one came up with this correction—if indeed the correction is correct—encourages me to think that a non-trivial point is at issue. It is a pleasure to record that final enlightenment came to me during conversations with my friend Al Moscowitz, which proves that chemical engineers can always benefit from consorting with physical chemists.

NOTATION

A	area of cross section
A*	reactant
a	radius of spherical particle
b	ratio of residence times in the two parts of the reactor [eq. (23)]
C	concentration of reactant in fluid phase or tubular reactor
$[C]$	average value of C over the whole reactor
C_f	feed concentration of reactant
c	concentration of reactant in solid phase or stirred tank
$\langle c \rangle$	average value of $c(r, t)$ over sphere
D	diffusion coefficient
D^*	dispersion coefficient in direction of flow
E	function defined by eq. (32)
F	function defined by eq. (9)
f	Laplace transform of F [eq. (11)]
G, H	functions defined by eq. (36)
k	reaction rate constant
k_c	mass transfer coefficient
L	length of reactor
$M, -N$	eigenvalues [eq. (36)]
m, n	dimensionless parameters [eq. (23)]
P	Peclet number for longitudinal dispersion
Q	parameter defined by eq. (39)
q	volumetric flow rate
r	radial distance within particle
S	area for interchange between tubular and stirred reactors
s	Laplace transform variable
t	time or age of particle
U	dimensionless concentration in tubular reactor
u	dimensionless concentration in stirred-tank reactor (also, following PVD, linear velocity)
V	volume
v	linear velocity of flow (also, following PVD, volumetric flow rate)
x	z/L
y	$Px = vz/D^*$
z	distance along the tubular reactor

Greek letters

α	dimensionless parameter defined by eq. (11)
β	dimensionless parameter defined by eq. (14)
Γ	function defined by eq. (40)
γ	dimensionless parameter defined by eq. (38)
Δ	function defined by eq. (40)
ε	fraction of bed occupied by fluid
η	effectiveness factor [eq. (38)]
θ	residence time, with suffix for particle, etc.
κ, μ, ν	dimensionless parameters defined in Appendix
ρ	r/a
σ	dimensionless Laplace variable
φ	Thiele modulus
ψ	dimensionless transfer parameter [eq. (38)]
χ	function of P defined by eq. (18)

REFERENCES

Amundson, N. R., 1946, Application of matrices and finite difference equations to binary distillation. *Trans. Am. Inst. chem. Engrs* **42**, 939.

Amundson, N. R., 1980, *The Mathematical Understanding of Chemical Engineering Systems*. Pergamon Press, Oxford.

Amundson, N. R. and Aris, R., 1962, Heat transfer in fluidized and moving beds, in *Proceedings of Symposium on Interactions between Fluids and Particles*, pp. 176–182. Institution of Chemical Engineers, London.

Aris, R., 1969, Canon and method in the arts and sciences. *Chem. Engng Educ.* **3**, 48.

Aris, R., 1977, Re, k and π. A conversation on some aspects of mathematical modelling. *Appl. math. Modelling* **1**, 386.

Aris, R., 1978, *Mathematical Modelling Techniques*. Pitmans, London.

Aris, R., 1979, De exemplo simulacrorum continuorum discretalumque. *Archs ration. Mech. Analysis* **70**, 203.

Aris, R., 1982, Chemical engineering in the university context. Hougen Lectures, University of Wisconsin.

Aris, R., 1983, The jail of shape. *Chem. Engng Commun.* **24**, 167.

Aris, R., 1989, Ut simulacrum poesis. *New Literary Hist.* **20**, 323.

Becker, H. A., 1976, *Dimensionless Parameters: Theory and Methodology*. John Wiley, New York.

Bridges, R., *The Testament of Beauty*, Book I; ll. 367, 8.

Casti, J. L., 1989, *Alternate Realities. Mathematical Models of Nature and Man*. John Wiley, New York.

Danckwerts, P. V., 1953, Continuous flow systems. Distribution of residence times. *Chem. Engng Sci.* **2**, 1.

Danckwerts, P. V., 1981, *Insights into Chemical Engineering*. Pergamon Press, Oxford.

Danckwerts, P. V., 1982, Review of "A Century of Chemical Engineering". *The chem. Engr* November, 406.

Denbigh, K. B., 1944, Velocity and yield in continuous reaction systems. *Trans. Faraday Soc.* **40**, 352.

Denbigh, K. B., 1947, Continuous reactions. The kinetics of steady state polymerization. *Trans. Faraday Soc.* **43**, 648.

Denbigh, K. B., Hicks, M. and Page, F. M., The kinetics of open reaction systems. *Trans. Faraday Soc.* **44**, 1948, 479.

Firth, J. D'E., 1936, *Winchester*, Chap. 12. Blackie, London.

Gilliland, E. R. and Mason, E. A., 1952, Gas mixing in beds of fluidized solids. *Ind. Engng Chem.* **44**, 218.

Iordache, O., 1987, *Polystochastic Models in Chemical Engineering*. VNU Science Press, Utrecht.

Layton, E. T., 1988, The dimensional revolution; new relations between theory and experiment in engineering in the age of Michelson, in *Proceedings of the Centennial Michelson–Morley Experiment*, pp. 23–38. Case Western University, Cleveland, OH.

Liljenroth, F. J., 1918, Starting and stability phenomena of ammonia—oxidation and similar reactions. *Chem. metall. Engng* **19**, 287.

Naor, P. and Shinnar, R., 1963, Representation and evaluation of residence time distributions. *Ind. Engng Chem. Fundam.* **2**, 278.

Nauman, E. B. and Buffham, B. A., 1983, *Mixing in Continuous Flow Systems*. John Wiley, New York.

Parulekar, S. J. and Ramkrishna, D., 1984, Tubular reactor stability revisited without the Danckwerts boundary conditions. *Chem. Engng Sci.* **39**, 455.

Penn, M. and Aris, R., 1980, The mere notion of a model. *Mathl Modelling* (*now Mathl Comput. Modelling*) **1**, 1.

Pethö, A. and Noble, R. D. (Eds), 1982, *Residence Time Distribution Theory in Chemical Engineering*. Verlag Chemie, Weinheim.

Sackville-West, V., 1934, The Land: Summer, in *Collected Poems*, pp. 86–87. New York.

Shinnar, R., 1987, Use of residence- and contact-time distributions in reactor design, in *Chemical Reaction and Reactor Engineering* (Edited by J. J. Carberry and A. Varma), Chap. 2. Marcel Dekker, New York.

Shuler, M. and Steinmeyer, D. E., 1989, A structured model for *Saccharomyces cerevisiae*. *Chem. Engng Sci.* **44**, 2017.

Smith, J. M., 1974, *Models in Ecology*. Cambridge University Press, Cambridge.

van Heerden, C., 1953, Autothermic processes. Properties and reactor design. *Ind. Engng Chem.* **45**, 1242.

Wagner, C., 1945, Über die Temperatureinstellung an Höchleistungskatalysatoren. *Chem. Tech.* **18**, 28.

Zwietering, Th. N., 1959, The degree of mixing in continuous flow systems. *Chem. Engng Sci.* **11**, 1.

ChE lecture

HOW TO GET THE MOST OUT OF AN EQUATION WITHOUT REALLY TRYING*

RUTHERFORD ARIS
University of Minnesota
Minneapolis, Minnesota 55455

MAXIMS FOR MATHEMATICAL MODELLING

1. Cast the problem in as elegant a form as possible.
2. Choose a sympathetic notation, but don't become too attached to it.
3. Make the variables dimensionless, since this is the only way in which their magnitudes take on general significance, but do not lose sight of the quantities which may have to be varied later on in the problem nor forget the physical origin of each part.
4. Use a priori bounds of physical or mathematical origin to keep all variables of the same order of magnitude, letting the dimensionless parameters show the relative size of the several terms.
5. Think geometrically. See when you can reduce the number of variables (even at the expense of first treating an over-simplified problem), but keep in mind the needs of the general case.
6. Use rough and ready methods, but don't carry them beyond their point of usefulness. (E.g. Isoclines in the phase plane).
7. Find critical points and how the system behaves near them or what is asymptotic behaviour is at long or short times.
8. Check limiting cases and see how they tie in with simpler problems that can be solved explicitly.
9. Use crude approximations, e.g. 1-point collocation. Trade on the analogies they suggest, but remember their limitations.
10. Rearrange the problem. Don't get fixed ideas on what are the knowns and what the unknows. Be prepared to work with implicit solutions.
11. Neglect small terms, but distinguish between regular and singular perturbations.
12. Use partial insights and despise them not. (E.g. Descarters' rule of signs).
13. These maxims will self-destruct. Make your own!

IT IS A COMMON PARADOX that one should only start computing after one knows the answer. Not to be taken to literally, it emphasizes that one should learn as much as possible about a problem before computing any case or sequence of cases so that the output of the computer may be critically appraised, for, without this critical oversight, the computer can produce an output more tedious and turgid than the so-called playboy philosophy. It is in any case part of the 'craft and sullen art' of the engineer or applied scientists to bring his problem into its most responsive formulation and to explore the modes of its solution as delicately as possible before proceeding to its complete analysis. From one point of view it requires sensibilities which are 'nascitur non fit,' but from another it is surely an art we may all strive after even if we despair of its mastery.

Of the texts on applied mathematics and engineering analysis the best may perhaps instruct by example, but only Segel and Lin's recent masterpiece [1] attempts to unfold some of the techniques of right formulation. There the question of reduction to dimensionless form and the scaling of equations is carefully and systematically explained. It will be clear that this essay is influenced by what they have done, both in this regard and in the play they have given to perturbation methods. The maxims of modelling that I have ventured to set down are a preliminary attempt to codify some of the mental processes of the chemical engineer as he probes and explores a problem. Like all maxims they tend to have the unassailable probity of "this ye ought to have done and not to have left the other undone."

***EDITOR'S NOTE: In this issue, *CEE* begins a new department: ChE LECTURES. We intend to publish seminars and lectures on important areas of modern chemical engineering. If you feel that one of your seminars or lectures on a certain topic would have pedagogical or tutorial value and would be of general interest to our erudite readers, please send the manuscript to the editor for review. We would appreciate comments from our readers on this new department as well as suggestions for authors of papers.**

Dr. Rutherford Aris was born in England in 1929, studied mathematics in the University of Edinburgh and taught it to engineers there. He has degrees from the University of London (B.Sc. (Math); Ph.D. (Math. and Chem. E.); D.Sc.). He worked a total of seven years in industry, but since 1958 he has been in the Chemical Engineering Department at the University of Minnesota enjoying the liveliness of its interests, both technical and cultural, and endeavouring to contribute to this vitality and communicate it to his students.

Nevertheless they should be looked on with a quizzical eye and subjected to a more than usually critical appraisal. They are dignified with numbers merely to invite the participation of the reader by pencilling them in the margin at the stage or stages of the example where they are most obviously invoked.

1. SETTING UP THE EQUATIONS

THE EXAMPLE WILL BE the elementary and familiar one of a single nonisothermal reaction in a catalyst pellet of arbitrary shape for, though it might be argued that I am getting the benefit of a good deal of hindsight, its very familiarity will allow us to concentrate on the method rather than be preoccupied with the matter. If the reaction is between the S species A_j it may be written $\Sigma\alpha_j A_j = 0$ giving a positive sign to the stoicheiometric coefficients of those species which are regarded as the products of the process. In the Knudsen diffusion regime the effective diffusion coefficients D_{ej} may be regarded as independent and mass balances for over an element of volume within the catalyst pellet for each species lead immediately to the S equations

$$D_{ej}\nabla^2 c_j + \alpha_j \rho_b S_g \check{r}(c_1,c_2,\ldots,c_S,T) = 0 \quad , \quad (1)$$

where $c_j = c_j(\mathbf{r})$ = concentration of A_j ,
T = temperature,
ρ_b = bulk density,
S_g = catalytic area per gram,
$\check{r}$ = reaction rate per unit catalytic area.

The Laplacian is with respect to the position variables $\mathbf{r} = (x,y,z)$ within the pellet which is assumed to have uniform properties. Into the formulation of this equation have gone the principle of the conversation of matter and two constitutive relations. One is a generalization of Fick's law which asserts that despite the physical complexity of the porous medium the flux can be related to the concentration gradient by an effective diffusion coefficient. The other is the kinetic law that may be embodied within the rate expression $\check{r}$. With the validity of the model we are not here concerned but though a suspension of disbelief is called for it should be remembered that it is ever temporary. An energy balance leads to the equation for the temperature

$$k_e\nabla^2 T + (-\Delta H)\rho_b S_g \check{r}(c_1,c_2,\ldots c_S,T) = 0 \quad (2)$$

where ΔH is the heat of reaction and is credited with a negative sign since the exothermic reaction, being more interesting in its effects, is taken as the norm. The simplest of boundary conditions will be taken at the boundary of the pellet, namely

The maxims of modelling that I have ventured to set down are a preliminary attempt to codify some of the mental processes of the chemical engineer, as he probes and explores a problem. Like all maxims they tend to have the unassailable probity of "this ye ought to have done and not to have left the other undone."

$$c_j = c_{jf}, \; T = T_f. \quad (3)$$

The notation for the basic equations is an obvious one with c_j immediately suggesting the concentration of the jth species and T the temperature. Similarly the suffix f in the boundary value suggests quantities associated with the fluid phase around the particle or, for the teutonically minded, with the surface. As a problem grows one often runs out of really sympathetic letters for the various quantities and compromises often have to be made. However, when the obvious suggestiveness of an initial letter (e.g. c,T) is abandoned, those quantities that hang together should have letters that hang together; thus dimensionless c and T may become u and v but the barbarity of Ψ and W should be avoided. Well-established conventions should be observed and of course there are publishers' house styles which may ultimately override a preference for Re and insist on N_{Re}. The practice of using two letters for one quantity is open to objection even though one in upper and one in lower case give it a pleasant literary favor.

However notation is somewhat a matter of taste and "de gustibus non est disputandum." Since it is also a vehicle of communication it is important not to become so attached to one's own version that the sensibilities of others are offended or communication impaired.

These basic equations presume a consistent set of units for each variable and parameter and our first task is to render the variables dimensionless. This does not derogate their physical significance in any way, for it is always important to keep the physical meaning of a variable or parameter in mind; rather it is intended to confer a meaning on their magnitude that is independent of the particular system of units. This point is important for the significance thus attained is universal in a deeper sense than would be conferred even by a universal agreement on units, such as the SI. Philosophically it is akin to Lonergan's independence of time and place [2]. But more than this, it measures the quantities in terms that are intrinsic to the problem rather than those dictated by an arbitrary external system. In general

However notation is somewhat a matter of taste and "de gustibus non as disputandum." Since it is also a vehicle of communication it is important not to become so attached to one's own version that the sensibilities of others are offended or communication impaired.

the objective should be to keep the dimensionless variables of the order of magnitude of 1 and allow the parameters to be just that—quantities which give the measure of the situation. However this should not be done at the expense of introducing unnecessary dimensions. For example, when the tubular reactor is considered without regard to any longitudinal dispersion there is no boundary at the far end of the reactor and it is artificial to introduce the length of the reactor just to make the dimensionless axial coordinate go from 0 to 1; it is preferable to use a combination of velocity and rate constant with the dimensions of length. It is however often possible to choose between putting a parameter in the equation or in the boundary conditions as we shall see later.

In the problem under consideration $\mathbf{r} = (x,y,z)$ is the coordinate system within the pellet of which we have some natural dimension, d_p, to render these independent variables dimensionless as $\rho = (\xi,\eta,\zeta) = \mathbf{r}/d_p = x/d_p, y/d_p, z/d_p)$. (Here a notational problem is raised by the traditional use of η for the effectiveness factor. We should probably go to (x_1,x_2,x_3) as coordinates with $\xi_i = x_i/d_p$. However so little use is made of the cartesian space coordinates that we will not belabour this. When there is symmetry and $\mathbf{r}$ can be taken as a scalar its dimensionless form is ρ.) It is dangerous to take $u_j = c_j/c_{jf}$ since we may want to consider $c_{jf} = 0$ for some products of the reaction. Rather we take c_f as characteristic of the c_{jf}, perhaps as $\Sigma_j c_{jf}$, and set $u_j = c_j/c_f$, $u_{jf} = c_{jf}/c_f$. Then equation (1) becomes

$$\nabla^2 u_j + \frac{\alpha_j d_p^2 \rho_b S_g}{D_{ej} c_f} \check{r}(c_f \mathbf{u}, T) = 0, \tag{4}$$

where the Laplacian is with respect to the dimensionless space variables. There is only one characteristic temperature in the data, namely T_f, and it is in no danger of being zero. We therefore take $v = T/T_f$ and the second equation is

$$\nabla^2 v + \frac{(-\Delta H)\, d_p^2 \rho_b S_g}{k_e T_f} \check{r}(c_f \mathbf{u}, T_f v) = 0. \tag{5}$$

The second terms in these equations is now dimensionless and could be written as say $R_j(\mathbf{u},v)$ and $R(\mathbf{u},v)$. However this would confound the importance of the various factors that enter the functions and overlook the fact that they are all proportional to one another. It is better to render the reaction rate dimensionless first by setting

$$R(\mathbf{u},v) = \check{r}(C_f \mathbf{u}, T)/\check{r}(c_{jf}, T_f) \tag{6}$$

so that $R(\mathbf{u}_f, 1) = 1$. Then the coefficient of R in equation (4) would be $\alpha_j d_p^2 \rho_b S_g \check{r}(c_{jf}, T_f)/D_{ej} c_f$ and this only depends on j through α_j and D_{ej}. If we let $\Delta_j = D_{ej}/D_e$ where D_e is some characteristic value of the diffusivities then $d_p^2 \rho_b S_g \check{r}(c_{jf}, T_f) D_e c_f$ emerges as the characteristic dimensionless parameter. We will not rush to fix the characteristic value D_e since it may be advantageous to fix it—or rather the combination $D_e c_f$—later. But

$$\phi^2 = d_p^2 \rho_b S_g \check{r}(c_f, T_f)/D_e c_f \tag{7}$$

. . . Our first task is to render the variables dimensionless. This does not derogate their physical significance in any way, for it is always important to keep the physical meaning of a variable or parameter in mind; rather it is intended to confer a meaning on their magnitude that is independent of the particular system of units. Philosophically it is akin to Lonergan's independence of time and place. But more than this, it measures the quantities in terms that are intrinsic to the problem rather than those dictated by an arbitrary external system.

is in fact the general form of the modulus introduced by Thiele and commonly bears his name. This is appropriate enough for of the three independent workers in the late thirties he solved this problem the most completely (see historical notes in 3, 4). Calling this parameter ϕ^2 we have

$$\nabla^2 u_j + (\alpha_j/\Delta_j)\phi^2 R(u,v) = 0 \quad (8)$$

$$\nabla^2 v + \beta\phi^2 R(u,v) = 0 \quad (9)$$

where $\beta = (-\Delta H) D_e c_f / k_e T_f$. These equations hold in Ω, the region occupied by the pellet, whilst on the boundary $\partial\Omega$

$$u_j = u_{jf}, \; v = 1. \quad (10)$$

Let us pause a moment to see what we have. There are three dimensionless independent variables in ρ occurring explicitly only in the differential operator, (S+1) dependent variables and one reaction rate expression, R. There are (S+2) visible parameters, of which S are stoicheiometric coefficients modified by the diffusion ratios (it is assumed that none of these is zero). β is clearly a dimensionless heat of reaction, on which more later, and there may be one or more parameters, such as a dimensionless activation energy or Arrhenius number, concealed in the dimensionless rate law. The Thiele modulus can be written

$$\phi^2 = d_p{}^3 \rho_b S_g \check{r}(c_{jf}, T_f) / d_p{}^2 D_e (c_f/d_p). \quad (10)$$

The numerator is proportional to the total reaction rate at surface conditions since the volume is some multiple of $d_p{}^3$. Similarly in the denominator $d_p{}^2$ is proportional to the surface area and (c_f/d_p) characteristic of the gradients in contration, so that the whole denominator is a measure of the total diffusion rate. The Thiele modulus is thus a ratio of the reaction rate to the diffusion rate and, when it is small, the reaction rate is the limiting whereas, when it is large, diffusion controls. But why call it ϕ^2 rather than ϕ? This is certainly legitimate since the Thiele modulus is always positive, but it is not altogether a product of hindsight. For the Laplacian is a second order operator and hence we might expect solutions to be functions of $\phi\rho$, which is rather neater than $\phi^{1/2}\rho$.

But at this point the mind should question whether (S+1) equations are really necessary when there is only one reaction in an adiabatic system. This is the import of the maxim "Think geometrically." Geometry is here being used in a sense which is loaded with even more overtones than the ἀγεωμέτρητος over the archway of the academy. It embraces the idea of degrees of freedom and of what we may expect in the way of characteristic dependencies. In this case we observe that we can eliminate the reaction rate between any pair of equations and that the linear combinations $(\beta\Delta_j u_j - \alpha_j v)$ are all harmonic functions. Since their values are constant on the surface $\partial\Omega$ they are constant throughout Ω and so each concentration can be expressed in terms of the temperature

$$u_j = u_{jf} + (\alpha_j/\beta\Delta_j)(v-1)$$

This would allow us to reduce all the equations to a single equation in the dimensionless temperature.

However, once having perceived the idea of reducing the equations to a single equation, we might try to do the reduction more symmetrically. There is a risk here since it is sometimes an advantage to stay with a physical variable such as v rather than move to a more mathematical variable. Let us brave this danger in the hope of comprehending the mysteries of the equation and set

$$u_j = u_{jf} + \frac{\alpha_j}{\Delta_j} w, \; v = 1 + \beta w \quad (11)$$

and

$$P(w) = R(u_{jf} + \alpha_j w/\Delta_j, 1 + \beta w). \quad (12)$$

Then all the equations reduce to the single equation

$$\begin{aligned} \nabla^2 w + \phi^2 P(w) &= 0 \text{ in } \Omega \\ w &= 0 \text{ on } \partial\Omega \end{aligned} \quad (13)$$

But $R(c,T)$ has a zero on the path

$$c_j = c_{jf} + D_e c_f \frac{\alpha_j}{D_{ej}} w, \quad T = T_f + \frac{(-\Delta H) D_e c_f}{k_e} w$$

for either the reaction reaches equilibrium or, if it is irreversible, a least abundant reactant is exhausted. We may now choose $D_e c_f$ so that this corresponds to $w = 1$. Then if the reaction is irreversible, $0 \leq w \leq 1$ since no concentration can become negative. If the reaction is reversible then $0 \leq w \leq 1$ provided the reaction does not go beyond equilibrium and the rate change sign. Physical intuition would dispose one to doubt if the reaction could go beyond equilibrium but the following argument (due, to Varma (5)) proves the case. For suppose there is a region Ω_- within Ω where $w>1$ and $P(w)<0$, then it is bounded by a surface $\partial\Omega_-$ on which $w = 1$. But in Ω_-, $\nabla^2 w = -\phi^2 P(w) > 0$ and so w is subharmonic. This implies that $w<1$ in Ω_- and contradicts the assumption that $w>1$ there.

We have reached the point of knowing that the system can be reduced to a single equation (13) in a variable w bounded between zero and 1; also the reaction rate expression $P(w)$ has been normalized so that $P(0) = 1$, $P(1) = 0$. By equation (11) the fact that $w \leq 1$ implies that the steady state temperature cannot exceed $T_f(1+\beta)$ a result which in its generality is due to Prater and justifies attaching his name to the parameter β. It was the way in which Thiele expressed the useful result of the solution of the equation that distinguished his work from that which had gone before. The mathematician commonly uses a norm of the solution, which in this case might be $\operatorname{Max}_{\Omega} |w|$, but the useful functional of the solution is not a norm but rather the average reaction rate as a fraction of the reaction rate at surface conditions. This is known as the effectiveness factor where $V_p = \upsilon d_p^3$ is the volume of the particle and $d\gamma = dV/d_p^3$ and $d\Sigma$ are the elements of volume of Ω and area of $\partial\Omega$ respectively. The external surface area of the particle is $S_x = \sigma d_p^2$ and $\partial/\partial n$ denotes the derivative along the outward normal. Note that for any shape η will be a function of ϕ, β and whatever parameters are concealed in $P(w)$.

$$\eta = \frac{\iiint \mathfrak{r}(c,T)\,dV}{V_p\, \mathfrak{r}(c_f,T_f)} = \frac{1}{\upsilon} \iiint_{\Omega} P(w(\rho))\,d\gamma = \frac{\sigma}{\upsilon\phi^2}\,\frac{1}{\sigma} \iint_{\partial\Omega} \left(-\frac{\partial w}{\partial n}\right) d\Sigma \qquad (14)$$

When the tubular reactor is considered without regard to any longitudinal dispersion there is no boundary at the far end of the reactor and it is artificial to introduce the length of the reactor just to make the dimensionless axial coordinate go from 0 to 1; it is preferable to use a combination of velocity and rate constant with the dimensions of length.

2. EXPLORING THE SOLUTION

THUS FAR WE HAVE only set up the equations that govern the system and it would be relatively safe to proceed immediately to the solution by some respectable numerical technique. We want, however, to get more of a feel for the form of the solution. To do this we can go in several directions:

a. simplify the geometry and with it the differential operator;
b. simplify the kinetics so that an analytical solution is possible;
c. use a crude numerical method;
d. consider limiting cases.

Let us consider these seriatim.

2a. SIMPLIFYING THE GEOMETRY

THE SIMPLEST FORM of a Laplacian operator is the second order derivative in one variable. To make the equation one dimensional we may consider the case of a slab of porous catalyst with two exposed faces a distance $2d_p$ apart and with its other edges sealed. Then there is a single spatial variable ρ, the dimensionless distance from the central plane and the exposed surfaces are $\rho = \pm 1$. To make things even simpler, we consider only symmetrical solutions for which the derivative vanishes on $\rho = 0$. Then the equations are:

$$\frac{d^2 w}{d\rho^2} = -\phi^2 P(w), \qquad (15)$$

$$\frac{dw}{d\rho} = 0, \rho = 0, \qquad (16)$$

$$w = 0, \rho = 1, \tag{17}$$

$$\eta = \int_0^1 P(w)\,d\rho = \frac{-1}{\phi^2}\left(\frac{dw}{d\rho}\right)_{\rho=1} \tag{18}$$

Though the sphere could have an equally symmetrical solution and is more natural, since it does not need sealed edges, its Laplacian is more complicated and ρ enters explicitly.

The second order autonomous form of equation (15) suggests the phase plane might give some insight into the solution. Let W denote the derivative $-dw/d\rho$, then

$$\frac{dw}{d\rho} = -W, \quad w(1) = 0, \tag{19}$$

$$\frac{dW}{d\rho} = +\phi^2 P(w), \quad W(0) = 0. \tag{20}$$

In the w, W plane this means that the trajectory $w(\rho)$, $W(\rho)$, $0 \leq \rho \leq 1$, is a curve, such as LM in Figure 1, which starts ($\rho = 0$) at some point on the w-axis and ends ($\rho = 1$) on the W-axis.

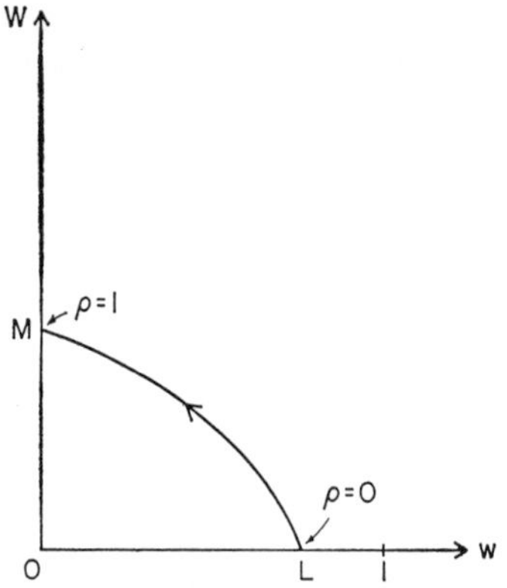

FIG. 1

The curve is a solution of the first-order nonlinear equation for W(w) which can be obtained by dividing (20) by (19), namely

$$\frac{dW}{dw} = -\phi^2\,\frac{P(w)}{W} \tag{21}$$

This equation can be solved for any value of ϕ and will give a solution if the path LM corresponds to going from $\rho = 0$ to $\rho = 1$ i.e. if

$$\int_L^M d\rho = \int_M^L \frac{dw}{W(w)} = 1.$$

The isoclines could be drawn in the plane and we could sketch the solution curves, but they would have to be redrawn for each value of ϕ. However ρ is only acting as a parameter along the solution curve and there is no reason why $\phi\rho$ should not be the parameter instead. Let $\phi\rho = \tau$, $s(\tau) = w(\tau/\phi)$, $S(\tau) = W(\tau/\phi)/\phi$ then

$$\frac{ds}{d\tau} = -S, \; s(\phi = 0, \tag{22}$$

$$\frac{dS}{d\tau} = P(s), \; S(0) = 0, \tag{23}$$

and

$$\eta = \frac{1}{\phi} S(\phi). \tag{24}$$

Now the isoclines can be drawn once and for all in the s, S plane, for suppose $dS/ds = -\omega$ then

$$\frac{dS}{ds} = -\frac{P(s)}{S} = -\omega \quad \text{or} \quad S = \frac{1}{\omega}P(s). \tag{25}$$

But this means the isoclines are all derived from the curve P(s) which represents the reaction rate expression and that for a given slope ω the curve for $\omega = 1$ is simply redrawn with a vertical scale of $1/\omega$. Let us suppose that the curve $S = P(s)$ is like PQ in Fig. 2. Then it can be crossed with a number of short lines of slope -1. The curve RQ whose ordinates are just twice those of PQ is ticked with lines of slope $-\frac{1}{2}$, whilst TQ at half the height of PQ is the isocline of slope -2. The s-axis and the vertical $s = 1$ correspond to infinite and zero slope respectively. Quite clearly then any solution curve such as LM will take off vertically from L and go in an arc of decreasing slope to M. In fact

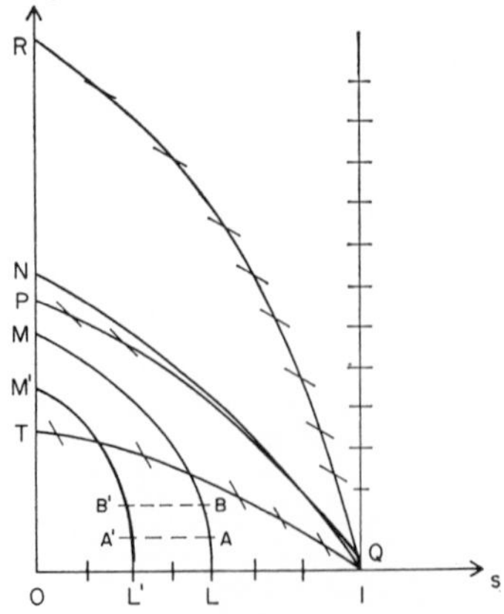

FIG. 2

if s_0 is the value of s at L, the arc S(s) is given by

$$S^2(s) = 2\int_S^{S_0} P(s')\,ds'$$

and M is the point

$$S^2 = 2\int_S^{S_0} P(s')\,ds' = [S(\phi)]^2. \qquad (26)$$

Once this curve is determined the value of ϕ which corresponds to it comes from the integral

$$\phi = \int_0^{\phi} d\tau = \int_S^{S_0} \frac{ds}{S(s)} \qquad (27)$$

Thus each trajectory can be made to yield a point on the η,ϕ-curve by equations (24), (26) and (27).

But what can we learn of the behavior of $\eta(\phi)$ without actually doing any of the integrations. First we see that for a solution curve L′M′ lying underneath LM the corresponding value of ϕ must be smaller. For the integral can be written

$$\phi = \int_0^{S(\phi)} \frac{dS}{-\omega(S)S}$$

and in comparing the contributions of the segments AB and A′B′ to their respective integrals we see that $(-\omega)$ is greater on A′B′ than on A B so that the integrand is smaller. Moreover the path LM is over a greater range of S than is L′M′ so that on both counts the value of ϕ corresponding to LM must be greater than that which corresponds to L′M′. The trajectories cannot cross one another (except at Q) hence a sequence of trajectories with increasing s_0 give an increasing sequence of values of ϕ.

When $\phi = 0$ the equation gives the solution $w = s = 0$, so that $s_0 = 0$ corresponds to $\phi = 0$. Does $s_0 = 1$ correspond to $\phi = \infty$? The answer must depend on the behavior of P(s) near its zero at $s = 1$. Let us suppose that $P(s) = -P'(1)(1\text{-}s) + 0(1\text{-}s)^2$ in the neighborhood of $s = 1$. Then the indeterminacy of dS/ds near $s = 1$, $S = 0$ is resolved by noting that

$$\frac{dS}{ds} = -\frac{P(s)}{S} = P'(1)\frac{(1-s)}{S}$$

can be integrated to give

$$S^2 = -P'(1)(1-s)^2.$$

The trajectory that starts from Q therefore takes off tangentially to the line $S = -\lambda(1-s)$ where $\lambda^2 = -P'(1)$. If this is substituted in the integral (27) with $s_0 = 1$ we see that the integral diverges. It follows that the trajectory through Q (QN in Fig. 2) does correspond to an infinite value of ϕ. What is remarkable however is that the solution curve QN does not go to infinity but reaches $s = 0$ for a finite value of N. This is clearly the case since if it sneaked up the S-axis it would have to have an increasingly large slope. But near the S-axis for large S the slope of the trajectories gets increasingly small, so that QN must finish at a finite point N. Let the value of S here be S_∞; then, since the trajectories LM move up under QN as ϕ increases, $S(\phi)$ approaches S_∞ as ϕ gets large. But equation (24) then shows that $\eta \sim S_\infty/\phi$ for large values of ϕ. Moreover the value of S_∞ can be calculated from equation (26),

$$S_\infty^2 = 2\int_0^1 P(s')\,ds'$$

Thus a rough sketch of isoclines can be made to yield a lot of information without really solving any equation. However, it should be mentioned that some of these arguments depend on the rather straightforward shape of P(w) and would not carry over quite so easily to a more general shape. In particular a family of nonintersecting curves of the type LM could be found, approaching 0 for $\phi \to 0$ and QN for $\phi \to \infty$ but they would not necessarily correspond to a monotonic sequence in ϕ. The arguments about QN would also have to be modified if $P'(1)$ were not finite.

2b. SIMPLIFYING THE KINETICS

THE ESSENTIAL NONLINEARITY of the equations lies in the kinetic expression P(w) which is limited only by the normalization $P(0) = 1$, $P(1) = 0$. If we take an isothermal ($\beta = 0$) first order reaction $P(w) = 1 - w$ and the equations become linear. In particular we have analytical solutions for simple shapes such as the sphere. In particular

$$\frac{1}{\rho^2}\frac{d}{d\rho}\left(\rho^2\frac{dw}{d\rho}\right)+\phi^2(1-w)=0 \qquad (28)$$

$$w(1)=0,\ w'(0)=0$$

has the solution

$$w(\rho)=1-\frac{\sinh\phi\rho}{\rho\sinh\phi} \qquad (29)$$

Thus the effectiveness factor is

$$\eta=\frac{3}{\phi}\left[\coth\phi-\frac{1}{\phi}\right] \qquad (30)$$

which again has the asymptotic property that $\eta\phi$ tends to a constant (in this case 3) as $\phi\to\infty$.

It is not suggested that the maxims or their illustration . . . provide an infallable recipe which when followed will open the portals of any problem. Rather, they are adumbrated as a framework.

In this case the complete solution can be obtained rather easily and one might use this as a starting point to explore other variations such as those of shape or in the boundary conditions. In any case it ties in with what we learned from the method of isoclines about the general behavior of the planar case. From the simple form of solution we see that the value of $w(\rho)$ rises from its zero boundary value with exponential sharpness near the surface $\rho=1$. In fact if $\rho=1-y$

$$w(\tau) \doteq 1 - \exp - \phi\, y$$

so that w quickly rises to a constant value of 1 when ϕ is large.

2c. COARSENING THE NUMERICAL METHOD

OFTEN SOMETHING CAN BE learned from an extremely crude numerical method. This was first shown for this problem by Stewart and Villadsen [6]. At any rate for small values of ϕ, $w(\rho)=\alpha(1-\rho^2)$ is an approximation that satisfies the boundary conditions for a symmetrical solution. We can deal with all three symmetrical shapes (the slab, cylinder and sphere) by writing the Laplacian as

$$\frac{1}{\rho^q}\frac{d}{d\rho}\left(\rho^q\frac{dw}{d\rho}\right)$$

and with this trial function for w it is

$$-2(q+1)\alpha=-2(q+1)w(\rho)/(1-\rho^2).$$

Thus the differential equation would be satisfied at the point $\rho=\rho_1$ if the value of α, and so of $w_1=w(\rho_1)=\alpha(1-\rho_1^2)$, were chosen to satisfy

$$2(q+1)w_1/(1-\rho_1^2)=\phi^2P(w_1). \qquad (31)$$

There is a full-scale theory, that of collocation methods, to say where the point ρ_1 is best taken, but we can use our experience with the sphere in the previous section. When $q=2$, for the sphere, and $P(w)=1-w$, for the first order reaction, equation (31) gives

$$w_1=\phi^2(1-\rho_1^2)/[6+\phi^2(1-\rho_1^2)].$$

Since the approximation should be good for small ϕ we might hope that this would agree with equation (29) for small ϕ. In fact the expansions are identical in the first term $\phi^2(1-\rho_1^2)/6$ and agree in the term of order ϕ^4 if $\rho_1^2=3/7$ or $\rho_1=0.6547$. A more general analysis would show that $\rho_1^2=(q+1)/(q+5)$ is a good choice.

Let us use this value of ρ_1 but return to a general $P(w)$. Then with $q=2$, $\rho_1^2=3/7$, equation (31) is

$$\frac{21w_1}{2\phi^2}=P(w_1), \qquad (32)$$

Before exploring this equation let us note that to the same approximation

$$-\left(\frac{dw}{d\rho}\right)_{\rho=1}=2\alpha=\frac{7w_1}{2}$$

Hence, by (14),

$$\eta=\frac{21w_1}{2\phi^2}=P(w_1). \qquad (33)$$

Equation (32) lends itself to a graphical solution, as is shown in Fig. 3, for the right hand side of the equation is the fixed curve $P(w)$ and the left side the straight line through 0 of slope $21/2\phi^2$. When ϕ is small the line is steeply sloped, like 0A, and η is close to 1. In fact, if the part of the curve $P(w)$ near $P(0)=1$ is approximated by the straight line $P(w)=1+P'(0)w$, then equation (33) can be solved for w_1 and

$$\eta = \left[1 - \frac{2P'(0)}{21}\phi^2\right]^{-1} = 1 + \frac{2P'(0)}{21}\phi^2 + 0(\phi^2) \quad (34)$$

On the other hand if ϕ is very large the same kind of straight line approximation gives

$$\eta = \frac{21}{2\phi^2}\left[1 + \frac{21}{2[-P'(1)]\phi^2}\right]^{-1}$$

This is not a very good approximation since we know that $\eta\phi$ tends to be constant for large ϕ. However this is not surprising since $w(\rho) = \alpha(1-\rho^2)$ is not a good approximation for large ϕ.

Much more important is the revelation of the possibility of multiple steady states that is made in Fig. 3. For if the curve $P(w)$ is as shown then for values of ϕ giving lines between OBC

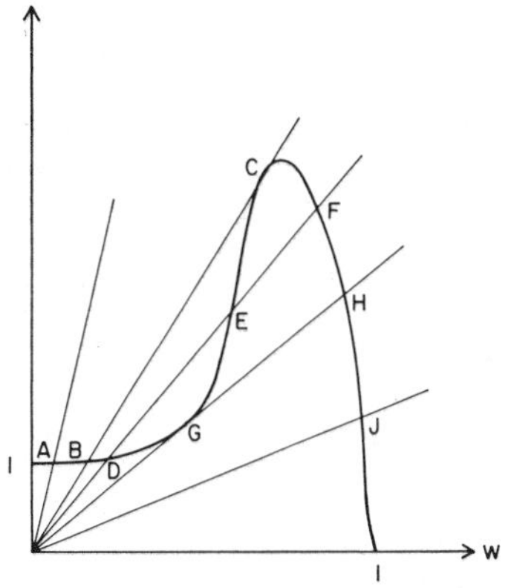

Fig. 3

and OGH there will be three intersections such as D, E and F which will give three values of η. We cannot expect much accuracy beyond the point G, but Stewart and Villadsen showed that a surprising accuracy was maintained [6]. The important thing is that it gives notice of the multiplicity of steady states. Furthermore the variable w_1 can be made a parameter in the computation of the η,ϕ-curve, since $\eta = P(w_1)$, $\phi^2 = 21w_1/2P(w_1)$. This suggests that some internal value of w, such as $w(0)$ or max $w(\rho)$ may be taken as the parameter along the η,ϕ-curve in more general cases.

On the basis of this understanding computation by more exact methods, such as those described by Villadsen and Stewart [6] and Finlayson [7], can safely go forward. At the same time one must be cautious not to push conclusions based on crude approximations too far.

2d. CONSIDERING LIMITING CASES

The limiting cases of large and small ϕ have already been considered in the partial solutions we have obtained. However they can be approached also from the equation itself. If $\phi = 0$, $\nabla^2 w = 0$ giving the solution $w = 0$ and $\eta = P(0) = 1$. Let us see if a solution can be generated in powers of ϕ^2 by setting

$$w(\rho) = \phi^2 w_1(\rho) + \phi^4 w_2(\rho) + \dots \quad .$$

The function $P(w)$ must be expanded similarly

$$P(w) = 1 + \phi^2 P'(0) w_1 + \phi^4[P'(0) w_2 + \tfrac{1}{2}P''(0) w_1^2] + \dots$$

Then comparing powers of ϕ^2 we have

$$\nabla^2 w_1 = -1 \quad \text{in } \Omega,\ w_1 = 0 \text{ on } \partial\Omega$$
$$\nabla^2 w_2 = -P'(0) w_1 \quad \text{in } \Omega,\ w_2 = 0 \text{ on } \partial\Omega, \dots \quad (35)$$

These equations are linear nonhomogeneous equations and are easier to solve than the nonlinear equation (13). Moreover we can see the form which η will take, for first w_1 can be found and averaged to give k_1 (say), then w_2 can be found and its average, k_2, calculated, and so on. This then gives

$$\eta = 1 + k_1 P'(0)\phi^2 + [k_2 P'(0) + \tfrac{1}{2}k_1^2 P''(0)]\phi^4 + \dots$$

which accords with equation (34).

This solution for small ϕ is obtained by a regular perturbation in powers of ϕ^2 and a regular perturbation does not involve a change in the character of the equations. The situation for large ϕ is quite different for, if we put $\epsilon^2 = 1/\phi^2$, we have

$$\epsilon^2 \nabla^2 w + P(w) = 0. \quad (36)$$

The limiting case $\epsilon = 0$ is here quite different for it has changed a second order differential equation into a non-differential equation $P(w) = 0$ whose only solution is $w = 1$. This makes sense since it claims that when ϕ is large (i.e. the reaction rate is vastly greater than the diffusion rate) the reaction is virtually complete everywhere. But it cannot be true near the surface, for $w = 0$ on the surface itself and the solution is a continuous function of position.

This is the classic situation of a singular perturbation problem in which an "inner" solution—in this case a solution near the boundary—has to be matched with an "outer" solution—in this case $w = 1$. Singular perturbation problems have a large literature [8,9,10,11] and this is no place to try to survey it, but mention should be made of the unusually lucid introduction that Segel and Lin [1] give in their book. In the present case we know from the experience of Section 2b that for large ϕ, w can rise with exponential sharpness from the boundary. In fact the solution we found there suggest that we should introduce ϕy where y is the normal distance from the boundary, as a new variable. This is known as a "stretching transformation" since it stretches y proportionately to ϕ. If we introduce an orthogonal coordinate system in the boundary surface, say ξ,η, and take $\zeta = \phi y$ as the third coordinate then the Laplacian in equation (36) is

$$\phi^2 \frac{\partial^2 w}{\partial \zeta^2} + \nabla_2^2 w$$

where ∇_2^2 is a second order operator in ξ and η. Substituting this in (36) and letting $\epsilon = 1/\phi$ tend to zero gives

$$\frac{d^2 w}{d\zeta^2} + P(w) = 0.$$

But this reduces the problem to the one dimensional case that we explored in Section 2a. Nor is this surprising since when all the change is confined to a thin shell on the outside of the pellet the curvature is not important and it might well be unfolded as a flat plate. Now the flat plate analysis gave

$$[S(\phi)]^2 = 2\int_0^{s_0} P(s')\,ds'$$

and when s_0 approaches 1, as in this case, $\phi \to \infty$ and

$$S \to S_\infty = [\,2\int_0^1 P(s')\,ds'\,]^{1/2}$$

But

$$-\frac{dw}{d\eta} = \frac{dw}{dy} = -\phi\frac{dw}{d\tau} = \phi S \sim \phi S_\infty$$

and so by equation (14)

$$\eta \sim \sigma S_\infty / \nu\phi \qquad (37)$$

This accords with all that has gone before. In particular if, for a sphere, d_p is the radius $\sigma = 4\pi$, $v = 4\pi/3$ while for a first order reaction $P(s') = 1 - s'$, $S_\infty = 1$. Thus $\eta \sim 3/\phi$ as we see also from equation (30).

This is the classic situation of a single perturbation problem in which an "inner" solution—in this case, a solution near the boundary —has to be matched with an "outer" solution . . .

CONCLUSION

The use of the phase plane and perturbation methods has been stressed in illustrating the value of the qualitative study of equations. These, of course, are not the only methods available—the maximum principles [12] and some of the theorems on the behavior of the solutions of equations come immediately to mind. Amundson's papers in general, and some of those with Luss [13] and Varma [14,15] in particular, show how the skillful use of such tools can give insight into much more complicated systems than the one considered here.

It is not suggested that the maxims, or their illustration in the above example, provide an infallible recipe which when followed will open the portals of any problem. Rather are they adumbrated as a framework within which one aspect of the craft of mathematical modelling may be

> **"Exercised in the still night**
> **When only the moon rages**
> **And the lovers lie abed**
> **With all their griefs in their arms."**

Sufficient will be the reward if for a few moments we find that it is "by singing light" that we have haply laboured. □

REFERENCES

1. Lin, C. C. and Segel, L. A. *Mathematics Applied to Deterministic Problems in the Natural Sciences.* Macmillan Pub. Co. New York 1974.
2. Lonergan, B. J. F. *Insight: A Study of Human Understanding.* Philosophical Library Inc. New York. 1958.
3. Aris, R. *Chem. Engng. Educ. 8*, 19 (1974).
4. Aris, R. *The Mathematical Theory of Diffusion and Reaction in Permeable Catalysts.* (2 vols.) Clarendon Press. Oxford. 1975.

5. Varma, A. *Chem. Engng. Sci. 29*, 1340 (1974).
6. Stewart, W. E. and Villadsen, *J. Chem. Engng. Sci. 22*, 1483 (1967).
7. Finlayson, B. A. *The Method of Weighted Residuals and Variational Principles*. Academic Press. New York 1972.
8. Acrivos, A. *Chem. Engng. Educ. 2*, 62 (1968).
9. Cole, J. D. *Perturbation Methods in Applied Mathematics*. Blaisdell. Waltham. 1968.
10. Murray, J. D. *Asymptotic Analysis*. Clarendon Press. Oxford. 1974.
11. Van Dyke, M. *Perturbation Methods in Fluid Mechanics*. Academic Press. New York. 1964.
12. Protter, M. H. and Weinberger, H. F. *Maximum Principles in Differential Equations*. Prentice-Hall. Englewood Cliffs. 1967.
13. Amundson, N. R. and Luss, D. *Can. J. Chem. Eng. 46*, 424 (1968).
14. Amundson, N. R. and Varma, A. *Can. J. Chem. Eng. 50*, 470 (1972).
15. Amundson, N. R. and Varma, A. *Can. J. Chem. Eng. 52* 580 (1974).

ACKNOWLEDGMENT

A preliminary version of this paper was given as a seminar in the UNESCO project of postgraduate education (VEN 31) at the Universidad Oriente, Puerto la Cruz, Venezuela under the local coordination of Prof. Hassan Elmayergi. It is a pleasure to record my indebtedness to Ray Fahien for many valuable conversations in which we probed the nature of the mathematician's "magnificent grasp of the obvious." I am also indebted to Professor Arvind Varma for some valuable comments.